HOW TO Rebuild and Modify Ford C4 AND C6 AUTOMATIC TRANSMISSIONS

George Reid

CarTech®

CarTech®

CarTech®, Inc.
6118 Main Street
North Branch, MN 55056
Phone: 651-277-1200 or 800-551-4754
Fax: 651-277-1203
www.cartechbooks.com

Edit by Scott Parkhurst
Layout by Monica Seiberlich

ISBN 978-1-934709-82-5
Item No. SA227

Library of Congress Cataloging-in-Publication Data

Reid, George.
How to rebuild & modify Ford C4 & C6 automatic transmissions / by George Reid.
p. cm.
ISBN 978-1-934709-82-5
1. Ford automobile--Motors--Maintenance and repair--Handbooks, manuals,
etc. 2. Automobiles--Transmission devices--Maintenance and repair--Handbooks, manuals, etc. I. Title. II. Title: How to rebuild and modify Ford C4 and C6 automatic transmissions.

TL215.F7R394 2012
629.28'72--dc23

2011044348

Written, edited, and designed in the U.S.A.
Printed in China
10 9

Title Page:
Make sure adjustment screws do not turn when you're tightening locknuts. Locknuts have integral seal lips to keep fluid inside.

Back Cover Photos

Top Left:
Pan and valve body are next, using a 1/2-inch socket for pan bolts and both 3/8- and 7/16-inch for valve body and filter.

Top Right:
Remove the intermediate band servo, checking cover and piston for any damage or abnormal wear. Remove the seals at this time. This is an "A" servo, which is the standard V-8 intermediate servo for compact and intermediate-size Fords. You're going to want an H or R servo, though it seems everyone wants the "C" 289 High-Performance servo. The H servo for big Fords and trucks is plentiful.

Middle Left:
The front pump, input shaft, and forward clutch come off next. With pump bolts removed, pump and forward clutch come right out.

Middle Right:
Ring gear and thrust washer are next to be removed. Examine each thrust washer for wear. Plan to replace all thrust washers and bushings.

Bottom Left:
Your performance rebuilding plan should include the 1967–1969 or 1970-on valve body (depending on which case you have) along with a proven shift-improvement kit.

Bottom Right:
Tighten in this sequence: intermediate band to 10 ft-lbs torque and then back off 1½ turns. Low-reverse band to10 ft-lbs and then back off 3 full turns. Locknuts to 35 to 45 ft-lbs. Many shops developed their own technique through the years, based on what works well for them, but this is what Ford suggests.

DISTRIBUTION BY:

Europe
PGUK
63 Hatton Garden
London EC1N 8LE, England
Phone: 020 7061 1980 • Fax: 020 7242 3725
www.pguk.co.uk

Australia
Renniks Publications Ltd.
3/37-39 Green Street
Banksmeadow, NSW 2109, Australia
Phone: 2 9695 7055 • Fax: 2 9695 7355
www.renniks.com

Canada
Login Canada
300 Saulteaux Crescent
Winnipeg, MB, R3J 3T2 Canada
Phone: 800 665 1148 • Fax: 800 665 0103
www.lb.ca

CONTENTS

Acknowledgments

This book has been one of the greatest challenges of my book-writing career because it has been much more involved than I ever imagined. When I accepted the invitation from CarTech to write this book, it seemed an easy task—build a couple of Ford automatic transmissions, have someone take pictures and good notes, and write about it. However, there's a lot more to an automatic transmission than meets the eye. A 3-speed automatic transmission is a complex series of components tied together and timed hydraulically to work smoothly and completely.

Getting an automatic transmission's components to work together has never been easy, not even for the most seasoned automatic transmission technician. In nearly half a century of time spent in auto repair shops, I've seen my share of professional builders stumped by chronic transmission problems they could not overcome.

Automatic transmission building is time consuming and demands exacting attention to detail. But transmission shops are never short on business, and that's why it was very difficult to get transmission shops to work with my photographers. Business is good because automatic transmission failure is quite common and customers always need their vehicles back right away.

The transmission-building professionals who agreed to work with me made many sacrifices to do so. They sidelined paying customers and made all kinds of time to ensure the success of this book. This is why it is important to recognize the five shops that helped make this book possible: Mike's Transmission, Tom's Transmissions, Leon's Transmissions, Transmission Rebuilding Company, and Mustangs Etc., all located in Southern California. The efforts of these talented transmission professionals and shop owners were enhanced with support from B&M Racing & Performance and TCI Automotive. These companies stepped up with kits, parts, transmissions, and technical support. Without their undying help and support, this book would not have been possible.

– George Reid
April 2012

What is a Workbench® Book?

This Workbench® Series book is the only book of its kind on the market. No other book offers the same combination of detailed hands-on information and revealing color photographs to illustrate transmission rebuilding. Rest assured, you have purchased an indispensable companion that will expertly guide you, one step at a time, through each important stage of the rebuilding process. This book is packed with real world techniques and practical tips for expertly performing rebuild procedures, not vague instructions or unnecessary processes. At-home mechanics or enthusiast builders strive for professional results, and the instruction in our Workbench® Series books help you realize pro-caliber results. Hundreds of photos guide you through the entire process from start to finish, with informative captions containing comprehensive instructions for every step of the process.

The step-by-step photo procedures also contain many additional photos that show how to install high-performance components, modify stock components for special applications, or even call attention to assembly steps that are critical to proper operation or safety. These are labeled with unique icons. These symbols represent an idea, and photos marked with the icons contain important, specialized information.

Here are some of the icons found in Workbench® books:

Important!– Calls special attention to a step or procedure, so that the procedure is correctly performed. This prevents damage to a vehicle, system, or component.

Save Money– Illustrates a method or alternate method of performing a rebuild step that will save money but still give acceptable results.

Torque Fasteners– Illustrates a fastener that must be properly tightened with a torque wrench at this point in the rebuild. The torque specs are usually provided in the step.

Special Tool– Illustrates the use of a special tool that may be required or can make the job easier (caption with photo explains further).

Performance Tip– Indicates a procedure or modification that can improve performance. Step most often applies to high-performance or racing engines.

Critical Inspection– Indicates that a component must be inspected to ensure proper operation of the engine.

Precision Measurement– Illustrates a precision measurement or adjustment that is required at this point in the rebuild.

Professional Mechanic Tip– Illustrates a step in the rebuild that non-professionals may not know. It may illustrate a shortcut, or a trick to improve reliability, prevent component damage, etc.

Documentation Required– Illustrates a point in the rebuild where the reader should write down a particular measurement, size, part number, etc. for later reference or photograph a part, area or system of the vehicle for future reference.

Tech Tip– Tech Tips provide brief coverage of important subject matter that doesn't naturally fall into the text or step-by-step procedures of a chapter. Tech Tips contain valuable hints, important info, or outstanding products that professionals have discovered after years of work. These will add to your understanding of the process, and help you get the most power, economy, and reliability from your engine.

CHAPTER 1

HISTORY AND FACTS

Ford had a fundamental challenge to its direction and future in the late 1950s—how to shed a stodgy image and dated technology. This effort began with a new generation of skirted-block FE-series V-8 engines in 1958. In 1960, Ford introduced its lightweight-iron Falcon and Comet sixes. The 90-degree Fairlane small-block V-8s followed in 1962. Prior to 1960, Ford cars and trucks were burdened with outdated, BorgWarner-designed cast-iron MX and FX automatic transmissions known as Ford-O-Matics, Merc-O-Matics, and Cruise-O-Matics. The MX was a large-case automatic and the FX was small. Although these transmissions were rugged and dependable, they were heavy, complex, and not easily adapted to performance applications. This is when Ford engineers developed lightweight aluminum-case automatic transmissions for an exciting lineup of automobiles that arrived in the 1960s.

When Ford Falcon and Mercury Comet were introduced for 1960, they were available with a new lightweight Ford-O-Matic 2-speed transmission. It was designed and manufactured by BorgWarner for new-generation gray-wall-iron straight-6 and small V-8s. What made the little Ford-O-Matic different than its predecessors was its aluminum case and steel hard parts inside and out. In early applications, the Ford-O-Matic transferred heat to the atmosphere via the torque converter and cooling vents in the bellhousing, instead of using fluid as coolant and a transmission cooler in the radiator. Later versions had a transmission fluid cooler in the radiator. The Ford-O-Matic and Merc-O-Matic were available behind the 144-, 170-, and 200-ci straight-6 engines, along with the 221- and 260-ci V-8s, which came later in 1962. The Ford-O-Matic/Merc-O-Matic had a case-fill dipstick tube. Bellhousing and main case were cast as one to reduce weight and reduce the likelihood of leakage. At first glance, the 2-speed automatic looks like a cast aluminum FX or MX case.

The 3-speed C4 Cruise-O-Matic was introduced just in time for the 1964 model year. The 1964–1966 C4 was known as the Dual-Range Cruise-O-Matic due to its dual-range shift pattern, which included two driving ranges based on shifter position. For 1967, the C4 went to a more conventional P-R-N-D-2-1 pattern and a different valve body. This C4 Dual-Range is an early V-8 unit with a five-bolt bellhousing. The B intermediate servo cover indicates mismatched parts because the B servo is for six-cylinder engines. Expect to see all kinds of mismatched transmissions.

The C4

Ford took what it learned from the 2-speed BorgWarner automatic and applied it to the C4 3-speed Cruise-O-Matic that arrived for the 1964 model year. The C4 was produced at Ford's Sharonville, Ohio, transmission plant for its entire production life through 1981 and was the first automatic transmission Ford designed and built. The C4 employed a new state-of-the-art Simpson compound planetary gear set, which became an industry standard in the years to follow. The C4 got its name from the model year that it entered production: "C" for the 1960s decade and "4" for the year, 1964. This naming practice didn't last long—witness transmissions to follow like the C3 in the 1970s and C5 in the 1980s.

C4 Gear Ratios

First Gear	2.46:1
Second Gear	1.46:1
Third Gear	1.00:1
Reverse Gear	2.20:1

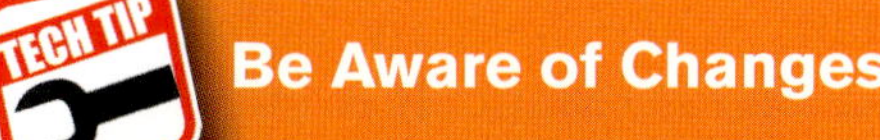

Be Aware of Changes

Watch out for running production changes in C4 transmissions. From 1964 to 1969, a 24/24-spline input shaft and forward clutch. In 1970 only, a 26/26-spline input shaft and forward clutch. And from 1971 on, a split-configuration 26/24-spline input shaft and forward clutch. ■

This is the C4 Dual-Range shift pattern for 1964–1966, with a large green dot for normal drive (1-2-3 upshifts) and a smaller dot for second gear only (driving on snow and ice). Too many motorists got this wrong and drove on the small dot, doing extensive transmission damage. For the 1966 model-year only, the C6 had this same shift pattern.

Beginning in 1967, Ford dropped the Dual-Range shift pattern for this traditional P-R-N-D-2-1 pattern and used a new name—Select-Shift. The shifter is placed at "2" for special driving conditions such as snow and ice, otherwise "D" for normal 1-2-3 upshifts. Both the C4 and C6 went to this shift pattern and name in 1967.

From 1964 to 1966 the C4 was called the Dual-Range Cruise-O-Matic—known among enthusiasts as the Green Dot transmission. The C4 Dual-Range is equipped with a valve body that allows a driver to start out in second gear on snow and ice with a 2-3 upshift, which is the small dot (off detent next to neutral) on the indicator. The larger green dot at the detent next to "L" enables you to start out in first gear and go through a normal 1-2-3 upshift program.

Ford called its C4 the Cruise-O-Matic while Mercury called its C4 the Merc-O-Matic. It is important to note "Cruise-O-Matic" was a broad marketing name that applied to Ford automatic transmissions of the mid-1960s era. Beginning in 1967, Cruise-O-Matic was dropped and the name "Select-Shift" was used for all Ford automatics.

For one model year only—1964—the C4 had a five-bolt bellhousing for V-8s only. In August 1964, the C4 and the V-8s it was mated to were fitted with a larger six-bolt bellhousing to reduce noise, vibration, and harshness.

For 1967, Ford did away with the Green Dot Dual-Range C4. Instead, it used a redesigned valve body offering a P-R-N-D-2-1 shift pattern known as Select-Shift. This valve body was used from 1967 to 1969. A redesigned C4 valve body and transmission case came along in 1970, which was used for the C4's production life through 1981. In Drive, the C4 shifts the same as the Dual-Range at the large green dot with a normal 1-2-3 upshift program and the same gear ratios throughout. If you want to start out in low gear on snow and ice or to creep along in slow traffic, all you have to do is place the shifter in "2" (second gear) for controlled starts and no upshift.

As the C4 evolved, other design changes were introduced. There were C4 transmissions with pan-fill dipstick tubes (blended case and bellhousing with 164-tooth flexplates). Most C4s were case-fill (notched case and bellhousing with 157-, or 148-tooth flexplates). Pan-fill C4 transmissions with 164-tooth flexplates and blended bellhousings were designed for full-size car and truck applications and are not recommended for compacts and intermediates because they just don't fit.

The 148-tooth bell and flexplate were designed specifically for Mustang II and Pinto/Bobcat/Capri with small transmission tunnels and are very hard to locate these days. There was also a version of the C4 produced with the 385-series (429/460) big-block bellhousing bolt pattern factory installed behind the 351M and 400M raised-deck Cleveland small-blocks in the 1970s. The 351/400M C4 is extremely rare because so few were produced.

The pan-fill C4 really is more about strength for heavy-duty applications like full-size cars and trucks than anything else because the bellhousing bolts to the case outside the pump housing. Case-fill C4 transmissions are light-duty; the bellhousing bolts to the front pump instead of the main case.

The 1964–1969 C4 input shaft and clutch hub size was .788 inch with a 24-spline on both ends. In 1970, Ford gave the C4 a larger input shaft and clutch hub measuring .839 inch with 26 splines on both ends for improved durability. From 1971 to 1982, the C4 had a split-spline count. It had a .839-inch input shaft with a 26/24-spline configuration, meaning a 26-spline at the torque converter and a 24-spline at the clutch hub.

C4 valve body variations are important to note because they're significant to your transmission-building project. At this time, I'm aware of at least four different types of C4 valve bodies and I suspect there are more out there. For 1964–1966, there's the Dual-Range/Green Dot valve body. At a glance, the Dual-Range valve body looks identical to 1967–1969. Internally, it has different valving and shift programming.

There's also the 1967–1969 valve body, which offers a conventional P-R-N-D-2-1 shift pattern.

For 1971–1981, the C4 valve body changed significantly and does not interchange with 1964–1969 bodies due to changes in the case. Case and valve-body bolt patterns changed for 1970–1981, which is why a 1964–1969 valve body does not fit a 1970–1981 case.

The round bell, six-bolt C6 transmission for FE-series big-block V-8s. A C6 is easily identified by its one-piece bellhousing and main case design.

Does it Interchange?

The C4's main case changed in 1970, from a nine-bolt valve body to an eight-bolt. Watch out for interchangeability issues here, between valve bodies and main cases. ■

The C5

In 1982 Ford introduced the C5 Select-Shift transmission, which was nothing more than a C4 with a locking torque converter to improve fuel economy. The C5 was in production between 1982 and 1986 at the Livonia, Michigan, transmission and axle plant and is not recommended as a performance transmission as received from the factory. However, C5 cases and many internal components are similar or identical to the C4, and are quite suitable for performance applications thanks to their improvements, as discussed in Chapter 4.

Like the C4, C5s were produced as both case-fill and pan-fill with either 157- or 164-tooth flexplates. None were 148-tooth flexplate. What makes the C5's main case desirable is improved oil circuits and some improvements to case strength.

Same, Yet Different

Architecturally, the C4 and C6 are very similar transmissions. What makes the C6 different is a low-reverse clutch package instead of a low-reverse band like a C4, making the C6 a single-band automatic. ■

The C6

Prior to 1966, Ford FE and MEL big-blocks were fitted with cast-iron MX and FX 3-speed automatic transmissions. For 1966, Ford introduced its own heavy-duty C6 3-speed automatic transmission for high-torque applications behind large-displacement big-block V-8s. Although the C6 has a completely different case and internal components than the C4, it is virtually the same internally to the C4—on a larger scale for heavy-duty use.

The C6 was produced with four basic bellhousing bolt patterns over its long production life and is a very rugged transmission designed for

C6 Gear Ratios

First	2.46:1
Second	1.46:1
Third	1.00:1
Reverse	2.00:1

Here's the C6 for 385-series 429/460 big-blocks as well as the 351M and 400M raised-deck small-block Cleveland V-8s, quickly identified by its finned case.

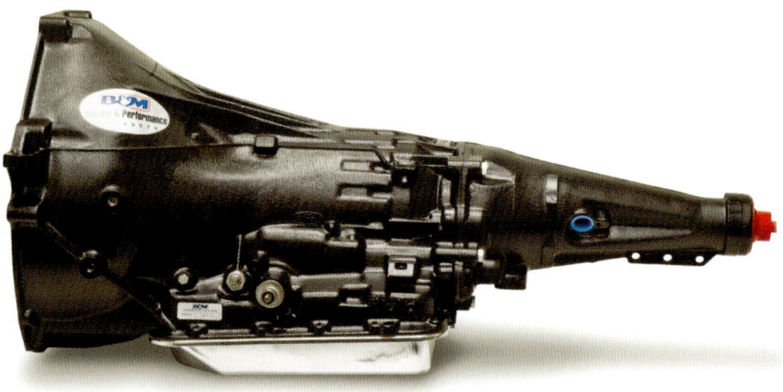

This B&M small-block C6 looks a lot like the big-block C6 unit except it's a small-block six-bolt bellhousing design. This case is also ribbed for strength.

high-power applications. The round six-bolt pattern is for FE-series big-blocks such as the 332, 352, 361, 360, 390, 406, 427, and 428 engines. There is another distinctive six-bolt bellhousing pattern for the 429/460-ci 385-series big-blocks and the 351M and 400M Cleveland-based, raised-deck V-8s. This six-bolt pattern arrived in 1968 with the 429/460 big-block V-8s.

There's also the small-block C6 originally intended for 351W and 351C engines, which fits any six-bolt 289/302/351W/351C small-block bellhousing bolt pattern.

Finally, there was a C6 for Diesel engines beginning in the 1980s, before the E4OD (4R100) was introduced in 1989, punctuating this transmission's reputation for durability. Despite the E4OD's presence, Ford continued to build the C6 until 1996 for industrial applications.

By the 1970s, Ford had a respectable lineup of modern lightweight automatic transmissions. An ironic footnote to this story is the weighty cast-iron FMX transmission, which remained in production until 1981 behind 351W small-block engines. It was an easy off-the-shelf solution for Ford, which needed the FMX to keep up with production demands when there weren't enough C4 and C6 transmissions to go around.

How to Read Ford Part and Casting Numbers

Ford part and casting numbers can be confusing. Once you come to understand this system, reading these numbers becomes second nature. There are actually two numbering systems. Here's how the Ford 1964–1996 part/casting numbering system looks:

C5AP - 7006 - A
(PREFIX - BASIC PART NO. - SUFFIX)

The prefix tells you when the part was originally released for production, what car line it was released for, and what engineering group it came from. The prefix breaks down like this:

First Position (Decade)

C = 1960–1969
E = 1980–1989
D = 1970–1979
F = 1990–1999

Second Position (Year of Decade)

4 = 1964
5 = 1965
6 = 1966
7 = 1967
8 = 1968
9 = 1969
0 = 1970

Then, the sequencing starts all over again at 1971 with "1," again in 1981 with "1," and again in 1991 at "1."

This C5 main case (RF-E2AP-7006-AA) is a 1982 casting. The C5 case is interchangeable with the C4 and has a better oiling circuit.

Casting numbers aren't always easy to read. This 1975 C4 case (D5OP-7006-AA) shows its vintage with a simple Ford casting number. Most cases have the Ford oval while others have "FoMoCo."

Valve bodies, like this one (D0AP-7A101-A-AU) for a C4, have casting numbers for quick identification. Remember, valve bodies are in two halves, which means casting numbers on both halves.

Third Position (Car Line)

A = Ford
D = Falcon
G = Comet, Montego, Cyclone
J = Marine and Industrial
K = Edsel
M = Mercury
O = Fairlane, Torino
S = Thunderbird
T = Ford Truck
V = Lincoln
W = Cougar
Z = Mustang

Basic Part Number

The basic part or casting number is the same whether it is an engineering number or a service number. For example, "7006" is the basic number for all automatic transmission main cases. What you're concerned with mostly here is the prefix, which tells you year and basic application.

Suffix

The suffix indicates the change level. "A" means original status of released part. "B" indicates at least one engineering change. The entire alphabet is used except the letters "I" and "L," which could be mistaken for the numeral 1. When Ford goes through the entire alphabet, it starts over again at AA, AB, AC, AD, and so on.

It is important to understand that part, casting, engineering, and service numbers rarely match each other. The casting number is derived from the actual casting or part, and typically does not match the part, engineering, or service numbers. Unless the casting has been revised, the basic casting number does not change. It means the number you see in the casting does not match the part number in the Ford Master Parts Catalog. And if the catalog you are using is dated, as most are, expect even more changes in your Ford dealer's microfiche or computer when it comes to suffixes. When demand for a part falls below a predetermined level, Ford discontinues or N/Rs the part. N/R means "Not Replaced."

Date coding works like this:

4D17
4 = 1964
D = April
17 = Day

If the date code is cast into the piece, it indicates the date the piece was cast at the foundry. If the date code is stamped or inked, it indicates date of manufacture.

CHAPTER 2

Getting Started

Before embarking on your Ford C4 or C6 transmission project, you're going to need to equip yourself with a proper work setting and tools necessary to do the job. Because automatic transmissions encompass dozens of tiny parts—clips, balls, pins, valves, springs, and other items—your workspace must be well-lit, neat, orderly. "Hospital clean" is an abused term, but it is appropriate in automatic transmission building. Even the smallest particle of dirt can disturb an automatic's precision tolerances, causing undesirable results and failure.

Safety Equipment

As in any other area of an automobile, automatic transmission building poses its share of safety hazards. You're going to be exposed to all kinds of toxic chemicals, which calls for hand, face, eye, lung, and ear protection. You're going to need nitrile rubber gloves, which are similar to latex types used in hospitals, to protect your hands from harsh chemicals. Solvents can dry your skin and pose a certain cancer risk. You also want to protect your skin from sharp edges. Iron and aluminum castings have their share of sharp edges, as do stamped sheet-metal components. Iron and aluminum particles can penetrate your skin and cause pain and infection.

High-frequency noise from power equipment can damage hearing, which calls for earplugs or muffs. Even the low-decibel din of shop equipment, electric motors, gear and belt drives, air compressors, and the like, damage hearing with exposure over time. Cleaning and drying parts with compressed air, which isn't always recommended, is loud enough to damage hearing.

Eye protection is paramount. Use goggles or safety glasses to protect your eyes from flying debris. Use

Safe shops are outfited with proper personal protection gear such as ear, eye, respiratory, and skin protection. Hearing loss can be directly attributed to loud noises, such as power tools and compressed air. Compressed air is probably the single greatest source of hearing loss and personal injury. Never use compressed air for horseplay.

wraparound goggles that keep debris well away from your eyes. Protect your face with a good face shield. Welding, for example, requires specialized eye protection to prevent permanent retina damage and sight loss.

Whenever you use harsh chemicals such as petroleum or alcohol-based cleaning solvents, protect your lungs with a good respirator. I see dust masks used occasionally—dust masks are not enough. A good respirator keeps all chemicals and particulates out of your lungs. The same rule applies to spraying paint. Always use a respirator, no matter how well ventilated your shop or driveway is.

Face Safety

Use face protection (a full-face shield) in addition to eye protection when working around power equipment. A bolt launched by a bench grinder can do permanent damage to your eyes and face. Metal particles from a hand-held grinder can do the same kind of damage. ■

A Place To Work

It is impossible to successfully rebuild an automatic transmission on your filthy garage floor or workbench, nor can you do it with an inadequate tool crib. You must have a clean, well-lit shop with a workbench. Dust and dirt are an automatic transmission's worst enemies. A tiny grain of dirt or sand damages or locks up an automatic transmission's precision parts.

Think of an automatic transmission like a Swiss watch—perfectly mated surfaces must have perfect tolerances. Clutch frictions and steels must mate precisely and smoothly. Sliding valves must glide through the valve body smoothly. Servo pistons and seals must be clean. When there's dirt, these parts bind because tolerances are extremely tight in order to provide proper containment of hydraulic pressure.

Seals can also be damaged by dirt and friction material, which causes internal hydraulic leaks and failure.

When you're not working on your transmission, keep it wrapped up tight inside a large heavy-duty plastic trash bag.

Your tool collection should include some type of transmission support fixture for both disassembly and build-up if you can afford it; it's much easier than building a transmission on your bench. Transmission shops typically have cobbled up holding fixtures made from transmission cases and shafts. There are also holding fixtures you can purchase from tool supply houses such as Harbor Freight, which offers cool shop tools for all kinds of operations at affordable prices.

You need an air compressor for air-tool functions, blowing out passages, and checking servo and clutch-piston function. Air leaks, when pressure is applied, tell you there are potential fluid leaks around servo and clutch-piston seals. The absence of a telltale click of a control-valve piston, when air pressure is applied, means something's amiss. Air tools also speed up the process of transmission disassembly and build-up. Good advice: always use a torque wrench (inch-pounds and foot-pounds) on all fasteners when it's time for assembly.

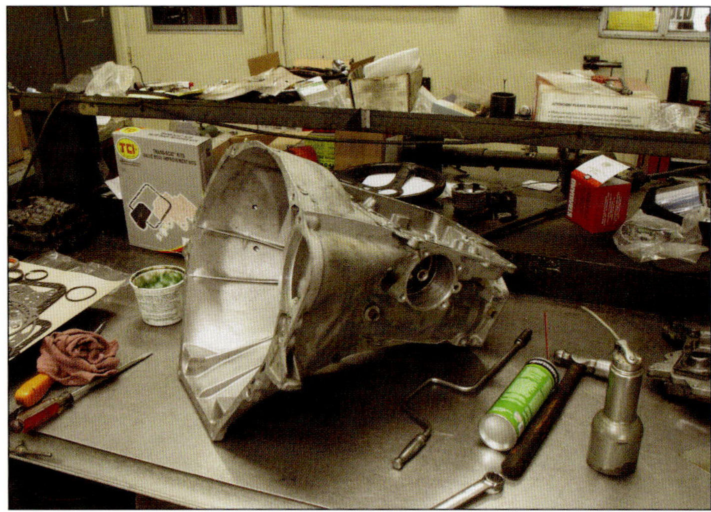

Regardless of the kind of automotive repair work you do, you must have a neat, clean place to work. Automatic transmissions are precision components that call for a clean working environment with a place for everything. You're going to need containers for small parts and you must have a hospital-clean work surface.

Tools of the Trade

There are tools in the automatic transmission trade you need to be familiar with: slide hammers, snap-ring pliers, mallets, tiny screwdrivers, awls and picks, presses, punches, bushing drivers, slide calipers, seal installation tools, roll-pin removal tools, alignment pins (made from old bolts), seal drivers and installers, Torx drivers, and more. Some of these items can be rented or borrowed, especially if you intend to do a transmission build only once. With others, you have to bite the bullet

Sears Craftsman combination wrench sets continue to have a lifetime guarantee, as do Snap-On and MAC tools, which makes them a good investment. When you invest in expensive tools, keep them in organizers like this one for good tool inventorying.

When it's time to wash parts, you're going to want a basket with holes smaller than your smallest parts. With automatic transmissions, this can be tricky because items like check balls have a way of slipping through.

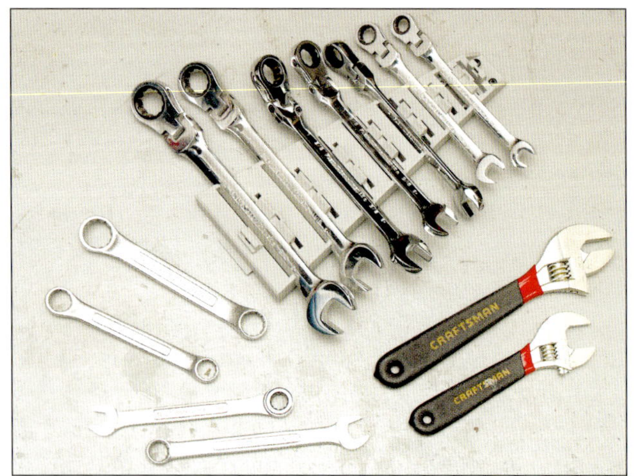

A good combination wrench set ranging in size from 1/4 through 1 inch should cover your needs. Avoid this kind of confusion and buy a matching combination wrench set with both SAE and metric sizes.

Harbor Freight is an outstanding bargain these days, with high-quality tools for not much money. Harbor Freight's Pittsburgh Pro Series tools are of good quality at bargain prices, which makes them a great value.

and buy them if you want to rebuild a transmission yourself.

Although automatic transmission building calls for specialized tools, most of the tools you're going to need are simple hand tools you can buy from Harbor Freight or Sears. Here's a good basic list of what you need:

- Combination Wrench Set (SAE), 1/4 to 1 inch
- 1/4-inch-drive Deep- and Shallow-Well Socket Set
- 3/8-inch-drive Deep- and Shallow-Well Socket Set
- 1/2-inch-drive Deep- and Shallow-Well Socket Set
- 3/8-inch-drive Speed Handle
- 3/8-inch-drive Breaker Bar
- All kinds of socket extensions of various lengths and sizes
- 3/8-inch-drive Torque Wrench (foot-pounds)
- 1/4-inch drive Torque Wrench (inch-pounds)
- Common and Phillips-Head Screwdrivers (all sizes)
- Awl and Pick Set
- Mallet
- Ball-peen Hammer
- Punch Set
- Tap and Die Set
- Thread Chaser
- Pliers
- Needle-nose Pliers
- Duck-bill Pliers
- Snap-ring Pliers Set
- Channel Locks
- Vise-Grip Pliers
- Putty Knife
- Wire Brush
- 4 Large C-Clamps (work just as well as a clutch pack assembly fixture)
- 55-Gallon Trash Bags (to use as dust covers)

Most professional transmission shops have transmission holding fixtures to support transmission cases during disassembly and assembly. Not many of us can afford such a fixture or even need it more than once. This calls for a hard work surface on which to build your C4 or C6. Although you might feel like you need a holding fixture, it really isn't necessary for the home garage technician. Where transmission disassembly and assembly gets tricky is when it is in the stack position and it's time to load components. I've seen some transmission shops use tailshaft housings as holding fixtures. Alternatively, you can always bore a hole in your workbench for a stack assembly. Places like Harbor Freight offer inexpensive holding fixtures, designed for most transmission types, that work well for disassembly and assembly.

Rebuilding any automatic transmission calls for compressed air. You need it for removing parts such as clutch pistons and sliding valves. You're also going to need it to clear passages. During assembly, you need compressed air to check component function. You need a huge industrial compressor; a 10- to 30-gallon compressor that operates off 110/115/120 vac produces the volume you need. And getting 220-vac power for more powerful compressors isn't always easy in some places.

Air tools make transmission building faster and easier. You're

You're going to need a variety of pliers ranging from conventional to needle-nose to duck-bill. Vise-Grip and channellock types are also important to your transmission-building tool inventory.

Snap-ring pliers are a must for transmission building—both internal and external types.

Snap-rings in automatic transmissions don't always have holes for snap-ring pliers. Make life easier by investing in convertible snap-ring pliers for all applications. You may also invest in a full-scale snap-ring pliers set such as the OTC 4512 set.

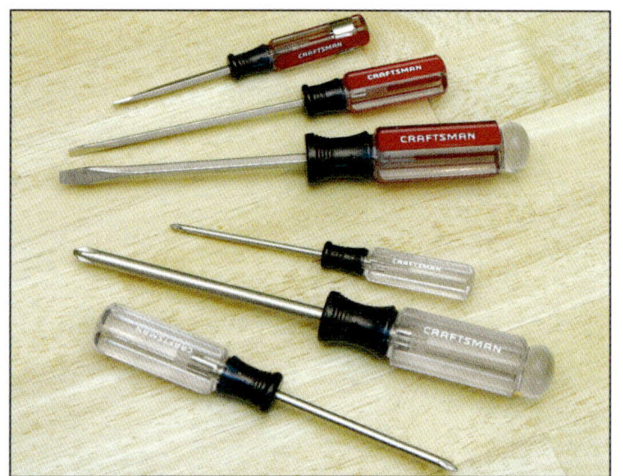

Sears offers the greatest selection of high-quality screwdrivers anywhere. A nice thing about Craftsman screwdrivers is color and feel, so you know by feel whether you're holding a common or a Phillips screwdriver. These screwdrivers are rugged and last a lifetime.

A speed handle makes light work of disassembly and assembly, enabling you to remove and install fasteners quickly.

Like the speed handle, air tools make tedious work go quickly. Never use an air tool to tighten fasteners. Use a calibrated torque wrench and be confident.

It's always good to have a bench grinder with a wire wheel for clean-up purposes. This is why eye protection is so important.

Rather than use a hammer like this one, go with a good ball-peen hammer and a 5-pound sledge. Know when to use a hammer and when to try something more gentle.

A tap-and-die set is a wise investment because it serves so many purposes. Transmission disassembly can involve damaged fasteners, pulled threads, and a lot of other problems where a tap-and-die set comes in handy. Invest in a good thread chaser set while you're at it, for clean threads and accurate torque readings.

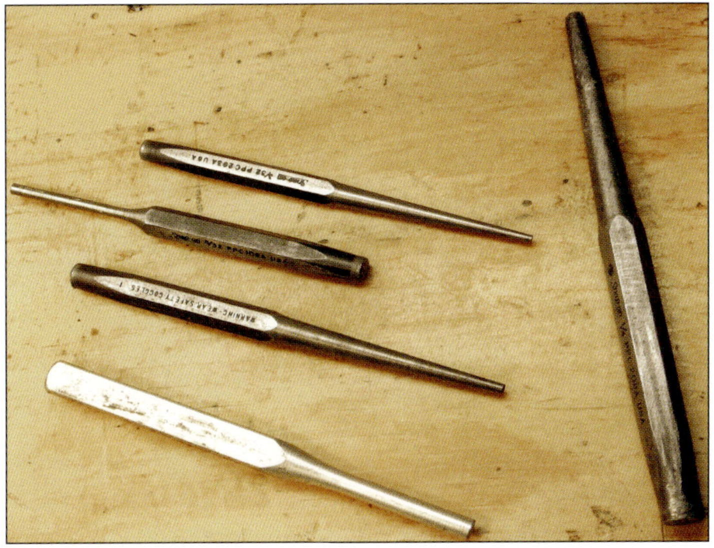

Transmission building calls for the use of punches from time to time. I've found the best punch set comes from Snap-On. Sears is the next-best option but only in high-carbon steel. Go with high-carbon steel punches, and wear eye and face protection.

going to need a 3/8-inch-drive air wrench and ratchet. A 1/2-inch-drive air ratchet is overkill for automatic transmission repair. If you're going to use an air grinder, take extra care not to damage cast aluminum contact surfaces.

An air blowgun is useful for clearing passages and for disassembling clutch packs and servos. It is also useful for doing operational checks on clutch packs, servos, and control valves. Always use eye, face, and ear protection whenever you use an air blowgun.

Because automatic transmissions have dozens of tiny parts, organization is very important. In my home shop, I use small, disposable kitchen containers and mark them for identification. Cooking sheets are also a good idea for larger parts like fasteners and can

Precision screwdrivers are something every home garage technician should have because they're good for tight spots. You may need them for valve body service issues and small seal removal.

I cannot overstress the need for precision measuring tools like calipers, thickness gauges, and the like for checking endplay and critical dimensions.

Invest in a complete set of high-quality drill bits and a drill bit sharpener. Few things frustrate more than a dull bit. And know the proper speed to drill—slow and easy along with lubrication is the best way. Have a couple of wire wheels on hand for clean-up purposes. Don't forget your eye protection.

This is a clutch spring compressor common to transmission shops. However, four C-clamps accomplish the same thing in your home shop.

be reused. Good old-fashioned coffee cans are good for cleaning parts. Lacquer thinner and brake cleaner are the best solvents because they have a high evaporation rate. Petroleum-based solvents are also good because they minimize the risk of rust and corrosion and make excellent grease cutters. When the heavy crud is gone, lacquer thinner and brake cleaner are good for final prep work.

For cleaning items like the main transmission case, tailshaft housing, and bellhousing, dishwashing detergent and a high-pressure washer work very well. A pressure washer can be rented or purchased for a modest amount of money.

Although hammers occupy most toolboxes, you're also going to want a hard mallet for things like servo covers and pistons. A mallet provides passive-aggressive force without inflicting injury. However, if you have to force any component, ask yourself why. Although servo covers and pistons must have a snug fit, installation by force means something's too tight or you forgot to lubricate seals and parts.

Picks and awls are handy for so many purposes, such as removing and installing C-clips or removing old O-rings.

A compressed-air nozzle is useful for blowing out passages and performing function tests. Compressed air enables you to check servo and clutch-piston function. You may also check control-valve function with compressed air.

This lip seal tool makes it easy and safe to roll in rubber seals without damage.

Fastener removal tools are very useful for stubborn fasteners seized to cases. Splash on plenty of WD-40 first.

Keep an oil can full of transmission fluid handy for lubricating parts and seals, and filling front pumps.

A set of seal-and-bushing drivers are affordable and every home garage should have one. Check out Harbor Freight for yours.

Although this is an older high-end torque wrench, it's a reminder that your garage should never be without one. You need both inch-pounds (3/8-inch drive) and foot-pounds (3/8- or 1/2-inch drive) for transmission work. Most of the time, you need inch-pounds—never get the two mixed up.

A universal transmission holding fixture is a nice touch for any garage and they don't have to be expensive. The more custom types get expensive.

A front pump puller is for just about any type of automatic transmission. But honestly, you don't need one. Most front pump assemblies pull with an aggressive yank. You may also work the front pump loose from inside your C4 or C6 with a large common screwdriver and caution.

Of course you're never going to have one of these in your garage, however, your local transmission shop should have one. A Hot Flush system (left) purges transmissions and coolers of stray particulates that can damage new transmissions. The Hot Flush system does a powerful two-way surge that dislodges clutch material and metal particles, catching all of it in the filter (right). This one belongs to Tom's Transmissions.

DISASSEMBLY

For the sake of simplicity and space I primarily cover a C4 teardown in this book. C4 teardown technique generally applies to the C6 as well, though there are some distinct differences in the two transmissions. You're going to be looking for the same elements of wear regardless of the type of vintage Ford automatic you're working with. Much of this may also be applied to the FX, MX, and FMX transmissions; the basic design and function isn't much different than the C4 or C6, aside from the planet package mostly, which is different on the FMX.

When automatic transmissions fail, there are physical reasons that can be identified when you pay close attention to details. Most of the time, transmissions fail from wear, tear, and neglect. As a matter of practice, motorists never service automatic transmissions until they do fail. However, transmissions need the same kind of attention engines do. They need regular fluid and filter changes if you want longevity.

Dirty fluid is a transmission's greatest enemy because it damages seals and causes excessive wear. And as fluid does its work, it deteriorates over time from heat and additive breakdown. As seals deteriorate, so does line pressure, which causes slippage and failure.

There are also irregularities in parts and assembly technique that can cause failure. Improperly machined or incompatible parts, dirt, or grit missed during assembly, and improper assembly of parts can cause failure. And, if missed a second or third time, failure is inevitable again unless you catch the problem this time. The buck stops with you, the transmission builder.

Transmission failure is normally a chain reaction—a series of small events and disruptions that ultimately lead to failure and a tow truck. The normal pattern of failure is at first seal wear/damage, which leads to weaknesses in hydraulic pressure, which leads to clutch and band slippage, which puts contaminants (friction material and metal) in the fluid. This damages seals further, hinders control pressure, and causes slippage and failure.

There are also types of mechanical failure, such as roller clutches and bushings, which fail with great regularity. When a roller clutch falls apart, you're not going anywhere. Clutches and bands burn up from slippage and extremes of stress such as towing or racing. Transmissions have an incredible job to do and must do it for thousands of miles, often without the benefit of service.

Transmissions also fail due to improper assembly technique in transmission shops. Whether it is the mass rebuilder or an independent shop, plenty gets overlooked in the course of a transmission teardown and rebuild, which leads to the same kind of failure again and again.

Study wear patterns and anything else out of the ordinary during teardown. Look for parts that haven't functioned well together, evidenced by scoring and other damage. Just because the transmission functioned normally doesn't mean it was properly assembled. Mass rebuilders, for example, toss parts in bins during the rebuild process with reckless abandon, clean them up, machine, or replace as necessary, and put transmissions together haphazardly. Transmission failure under warranty is considered part of the cost of doing business. As a matter of economics, rebuilders reuse hard parts that should be machined, rebuilt, or tossed in the recycle bin so they can never threaten another transmission's lifespan. Carelessness also means generations of transmission parts thrown together that aren't always designed to work together.

Ford is infamous for annoying engineering changes, dozens of them over the production life of a unit design, which means you must pay very close attention to parts intended to work together and parts that were never intended to be together. The C4, for example, had at least three major production changes between 1964 and 1971 that are very significant, which means you can wind up with parts that don't even fit, let alone work well together. Also, changes came well after 1971 as durability requirements increased in the late 1970s.

Disassembly is an opportunity to not only learn what went wrong with a transmission, but to also learn how it should go back together. Take lots of pictures and notes as you go along. And when something just doesn't look right, take note of it and don't be afraid to ask questions. There are plenty of resources for odd information and production/engineering changes via the Ford Master Parts Catalog and Transtar Industries. Aftermarket companies like TCI Automotive, B&M Racing & Performance, Performance Automatic, and Trans-Go are excellent sources for information and parts.

C4 Disassembly

1 Remove Servo and Piston

Disassembly begins with external parts like the neutral safety switch, kickdown linkage, and any brackets. First remove the low-reverse servo cover and piston (C4 and C5 only). Inspect the piston for damage and a worn seal. Replace the low-reverse piston in any case.

2 Remove Throttle Valve

Remove the throttle valve (vacuum modulator), including the valve body control rod (pin) inside the case. C4 transmissions from 1964 to 1971 have screw-in throttle valves. From 1972-on, expect to see a pressed-in throttle valve with retaining bracket. The same can be said for a C6 with a screw-in throttle valve from 1966 to 1971, and then press-in from 1972-on.

3 Remove Pan & Valve Body

Pan and valve body are next, using a 1/2-inch socket for pan bolts and both 3/8- and 7/16-inch for valve body and filter.

4 Remove Tailshaft Housing

The tailshaft housing comes off next, using a 9/16-inch deep-well socket.

Important!

5 Remove Governor

Remove the the governor and check for freedom of valve movement. Both primary and secondary valves should move freely, detected by a rattle. There should also be a tiny filtration screen in the tailshaft governor flange. But be prepared; most are gone and you may need to replace yours. See Transtar or your local transmission parts supplier for this screen.

Critical Inspection

6 Inspect Governor

Governor assembly has two points of inspection: primary and secondary control-valve pistons for freedom of movement and spring integrity. It's a good idea to disassemble the governor and do an inspection, taking note of how it comes apart.

7 Remove Servo

Remove the intermediate band servo, checking cover and piston for any damage or abnormal wear. Remove the seals at this time. This is an "A" servo, which is the standard V-8 intermediate servo for compact and intermediate-size Fords. You're going to want an H or R servo, though it seems everyone wants the "C" 289 High Performance servo. The H servo for big Fords and trucks is plentiful.

8 Remove Band Struts

Remove and inspect intermediate and low-reverse band struts. The C6 (not pictured here) doesn't have a low-reverse band, but instead a low-reverse clutch package, which is part of the main case.

9 Inspect Geartrain

With band struts removed, you can see the C4's geartrain package.

Internal Disassembly

1 Remove Pump, Input Shaft & Forward Clutch

The front pump, input shaft, and forward clutch come off next. With pump bolts removed, pump and forward clutch come right out.

2 Remove Intermediate Band

After forward clutch and pump are out, remove the intermediate band.

Critical Inspection

3 Inspect Intermediate Band

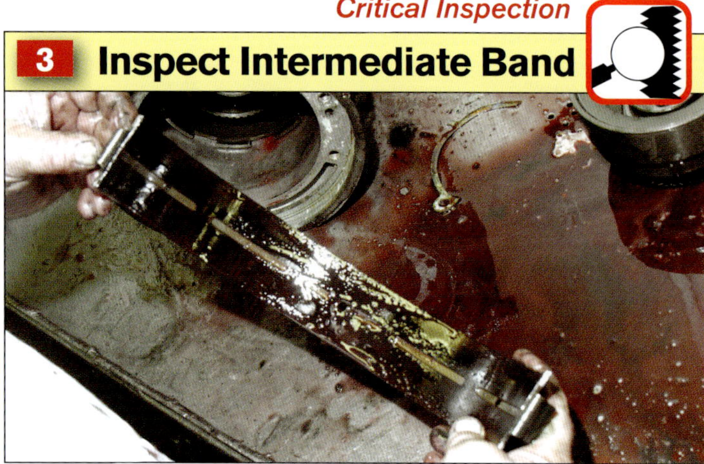

Inspect the intermediate band for worn and burned friction surfaces. A burned intermediate band is a sign of poor adjustment or inadequate servo pressure. It's a good idea to replace the intermediate band regardless of condition. Be choosy about replacement band quality; this is no place to cut corners.

Critical Inspection

4 Disassemble Front Pump

Next, disassemble the front pump and inspect for abnormal wear patterns. Mating surfaces should be smooth and void of scoring. Excessively worn gear and pump-cavity surfaces are grounds for replacement or machine work. Scoring adversely affects pump pressure. Inspect pressure relief valves for sticking, proper spring pressure, and any debris.

5 Remove Gear Shell

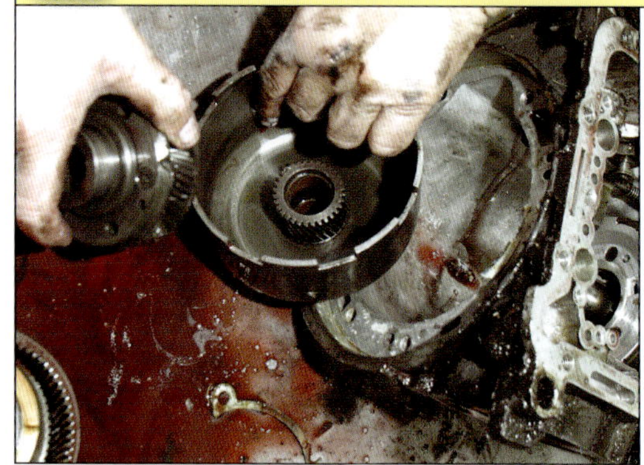

Remove the Sun gear shell, forward planet, and ring gear and inspect for wear and damage.

6 Remove C-Clip

Remove the tailshaft C-clip next, which releases the tailshaft.

7 Remove Governor Support

Remove the governor support assembly, along with two pressed-in hydraulic lines.

8 Remove Parking Pawl

Remove the parking pawl. Inspect gear, pawl, and thrust washer for damage and wear. Anything marginal (cracks, chaffing, wear) should be replaced.

9 Remove Ring Gear & Thrust Washer

Ring gear and thrust washer are next to be removed. Examine each thrust washer for wear. Plan to replace all thrust washers and bushings.

Critical Inspection

10 Inspect Forward Planet Carrier

Forward planet carrier does most of your C4's work during acceleration and upshift. Do a complete inspection, checking planet gears for excess wear and oscillation. Operation should be smooth and without oscillation. Any resistance to rotation (binding) is reason for discard or rebuild. As a rule, this is not a serviceable piece.

Critical Inspection

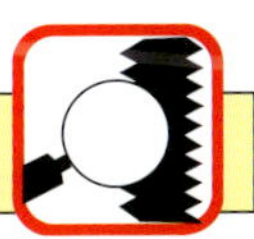

11 Inspect Forward Planet Torrington Bearing

Inspect forward planet Torrington bearing for wear. Most of the time, it's a good idea to replace this thrust while you're in there.

Critical Inspection

12 Inspect Rear Planet Carrier

Rear planet carrier gets less wear and tear. However, it calls for the same kind of detailed inspection as the forward planet, checking planet gears for excessive wear and oscillation and the carrier for cracks and other damage.

13 Remove Tailshaft Sealing Rings

At this time remove the tailshaft/governor support sealing rings. These are not gapless rings as found elsewhere in a C4 transmission. Always replace sealing rings in any case.

14 Remove Front Pump Sealing Rings

Front pump/forward clutch sealing rings are gapless types. They should also be removed and replaced.

15 Remove Sealing Rings

Remove forward clutch/front pump sealing ring as it sits inside forward clutch. Contact surfaces should be free of scoring and excessive wear. Any damage calls for machining or replacement.

Clutch/Band Disassembly

1 Disassemble Reverse-High Clutch Assembly

Disassemble the reverse-high clutch assembly. This is an early C4 reverse-high clutch with 10 return springs. Later models have one large return spring. Piston seals are removed at this time. Check clutch piston for wear and check ball integrity.

Critical Inspection

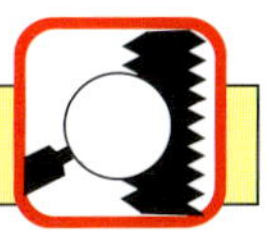

2 Inspect Reverse-High Clutches

Inspect reverse-high clutches for wear and evidence of slippage and burning. Also inspect clutch drive plates for slippage and heat damage. Heat damage shows up as brown, black, or blue score marks. If drive plates look undamaged, all you have to do is dress them with fine sandpaper (280-grit). If they have heat damage, they must be replaced. Not all transmission builders agree with the idea of resurfacing drive plates (steels) because they don't come out of the box machined to begin with. The same can be said for clutch drums.

3 Disassemble Forward Clutch

Next, disassemble the forward clutch by removing perimeter snap-ring, clutches, and drive plates.

4 Remove Belleville Spring

The forward clutch piston is returned to rest by a Belleville spring, also known as a disc spring. A snap-ring retains the C4's Belleville spring.

Critical Inspection

5 Inspect Low-Reverse Band Servo

Inspect the low-reverse band servo piston for seal damage. It is a good idea to replace the low-reverse band servo piston as a matter of practice regardless of condition because the seal is probably hard and worn from heat. Seal and piston are bonded into one assembly. Seal leakage (loss in pressure) leads to transmission slippage and failure.

Critical Inspection

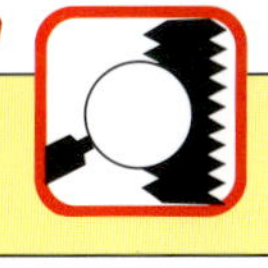

6 Inspect Intermediate Band Servo

Inspect the intermediate band servo piston for wear and check-ball function. Seals must always be replaced. Take note of the type of seals and how they're installed during disassembly. Some builders suggest keeping the old seals installed during disassembly and cleanup to reduce the risk of installing incorrect seals or installing seals incorrectly.

7 Remove Low-Reverse Band and High-Reverse Clutch Drum

Remove the low-reverse band and reverse-high clutch drum and inspect for wear, heat damage, and scoring. As a rule, low-reverse bands are pretty hardy and hold up quite well. The best low-reverse band to buy is new-old-stock Ford or OEM if you can find it, which offers the better friction materials than you see today from the aftermarket. When new-old-stock isn't realistic, go with the best low-reverse band money can buy. Transtar or Trans-Go are among the best sources.

Documentation Required

8 Remove One-Way Clutch

Take note of how the rollers and springs are installed when you remove the one-way clutch. Take pictures of this assembly prior to disassembly. It is easy to improperly reinstall rollers and springs. Be extra careful here. The roller clutch's job is to allow low-reverse drum rotation one way, but not the other.

9 Remove Manual Shift

Next, remove the manual shift and kickdown linkages by removing the retaining nut inside with a 7/8-inch open-end wrench. This frees up the manual and kickdown linkages. Take note of how these "shaft within a shaft" levers are installed. There's an outer manual-shift-lever lip seal and an inner kickdown lever O-ring seal to remove at this time.

Professional Mechanic Tip

10 Use Flat-Blade Screwdriver

Manual shift lever lip seal pops out with a large flat-blade screwdriver.

Critical Inspection

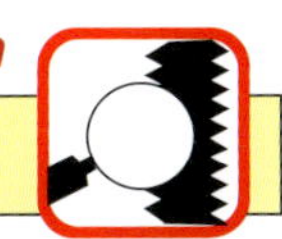

11 Inspect Ring Gears

Inspect ring gears for abnormal wear. Leon's Transmissions strongly suggests disassembling ring gear assemblies because metal and stray friction material gets trapped here, which can later cause failure.

Critical Inspection

12 Inspect Thrust Washers

Closely inspect the thrust washers, which provide a load-bearing surface between rotating geartrain parts. Measure thicknesses and closely examine these washers for wear. They can be reused if no real wear is found.

The front pump has a nylon thrust washer (number-2), which you should replace if it exhibits heat damage. This one is very discolored from excessive heat damage and should be replaced. Washers number-1 and -2 determine geartrain endplay, which makes them known as "selective" washers because they come in various thicknesses to help set proper geartrain endplay as it relates to the front pump.

Professional Mechanic Tip

13 Remove C-Clip

Ford's C4 and C6 transmissions are among the easiest in the industry to rebuild in your home garage. Transmission shops have special spring compressors for clutch disassembly; however, you can do this at home with four C-clamps. This makes it easy to remove the C-clip.

14 Remove Clutch Pistons

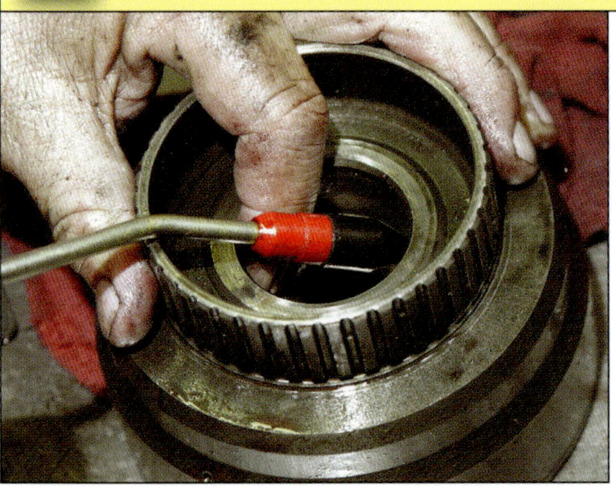

Use compressed air to remove clutch pistons. Turn clutch drum face down for this procedure to protect your eyes and face. Piston should pop right out.

15 Note Any Clutch Wear

Clutch wear is a strong indicator of transmission health. When clutches are badly burned from slippage, ask yourself why there was slippage to begin with. The same can be said for band heat damage. Ask yourself why there was slippage and be diligent in finding the cause. Most of the time, slippage is caused by insufficient line pressure at clutches and bands. Surface irregularities can also cause slippage and heat. Hot spots indicate high and low spots—irregularities in surfaces. When in doubt, replace. These steels have excessive scoring and should be replaced.

16 Use Compatible Parts

If you have to replace the input shaft or forward clutch, make sure you're getting compatible parts. There are three basic scenarios: 1964–1969, 24-spline both ends; 1970-only, 26/26-spline both ends; and 1971-on, 24/26-split-spline. The best-case scenario is 26/26 if you can find it, which was originally 1970 only. However, the aftermarket can help you with a 26/26 shaft and forward clutch.

C6 Teardown Notebook

Ford took what it learned from C4 development and production and applied it to the new heavy-duty C6 transmission first introduced for the 1966 model year. The C6 was a lighter-weight replacement for those older FX and MX iron heavyweights used behind FE and MEL big-blocks and Y-blocks for so many years.

What makes the C6 different than a C4 is its one-piece bellhousing and main case design, plus a low-reverse clutch instead of a low-reverse band and drum. The low-reverse clutch package makes the C6 more rugged and certainly more reliable, making the C6 among the most durable automatic transmissions ever produced.

Ford then took what it learned from the C6 and applied it to the heavy-duty E4OD overdrive transmission that came in 1989. Inside every E4OD are modified and improved C6 internals.

Beyond the E4OD has been the 4R100, which is basically an E4OD. Unfortunately, little if anything from the 4R100 fits your C6.

C6 Disassembly

Brian Fortune, of Tom's Transmissions, performs the following C6 teardown and buildup. Along with TCI Automotive he shows you how to build a better-performing C6 transmission using E4OD geartrain components, which provide durability and better gearing for improved acceleration. I don't kid myself about the C6 though; it is not a drag racing transmission because it is too heavy and eats up a lot of power. For tow vehicles and classic Fords, it is an outstanding piece because it was so rugged.

C6 Disassembly

1 Prepare For Disassembly

Although the heavy-duty C6 transmission has similar architecture as the C4, it does have differences that need to be addressed. Bellhousing and main case are a one-piece design. Instead of a low-reverse clutch drum and band, you have a low-reverse clutch tied to the case. This is a mid-1970s C6 transmission.

2 Note Bolt Locations

C6 valve body removal involves 3/8- and 5/16-inch sockets. Take note of bolt locations.

Documentation Required

3 Remove Servo

Like the C4, the C6 has an intermediate servo and band. Servo removal is the same as C4 removal using a 1/2-inch socket. Remove servo piston seals and take note of what kind of seals your servo piston has. The most popular C6 intermediate servo is the R servo (the largest) which provides solid band hookup.

4 Remove Pump

C6 front pump removal is carried out the same way as C4, including input shaft removal. When the pump is out remove the forward clutch.

5 Remove Forward Clutch

Forward clutch assembly comes out just like the C4's and appears identical except for its larger size.

Critical Inspection

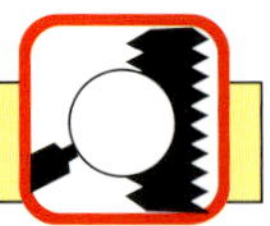

6 Remove Forward Planet Carrier

Forward planet carrier from this mid-1970s C6 is aluminum for weight reduction. Expect to see iron or aluminum. As with the C4's forward planet, inspect planet gears for wear and damage. There should be no oscillation around each planet gear's centerline or resistance to being turned. Each gear should roll smoothly without sideplay or much endplay. Inspect the carrier for cracks or damage.

7 Remove Rear Planet Carrier

C6 rear planet carrier works exactly like the C4's rear planet, which fits right into the sun gear shell. And like the C4's rear planet, this is also an aluminum carrier.

8 Remove Tailshaft C-Clip

The tailshaft is secured with a C-clip. Remove the C-clip and free the tailshaft.

9 Remove Low-Reverse Hug, Ring Gear & Roller Clutch

After the tailshaft is free remove the low-reverse clutch hub, ring gear, and roller clutch. Ring gear fits inside the low-reverse clutch hub.

10 Remove Low-Reverse Clutch

With snap-ring remove the low-reverse clutches and drive plates. The low-reverse clutch package does the same thing the low-reverse drum and band do in a C4. Clutches are quicker and more durable than bands and drums.

11 Note Parking Pawl

Parking pawl and gear on the C6 is similar to those in the C4—quite visible when the governor support is removed.

Documentation Required

12 Remove Roller Clutch

Free the one-way roller clutch with a 7/16-inch socket. Again, take note of roller and spring configuration.

Professional Mechanic Tip

13 Inspect Reverse-High Clutch

Examine the reverse-high clutches for wear and replace if indicated. As with the C4, examine friction and steel surfaces for heat damage. Drive plates should be resurfaced to improve hook-up, or replaced if there's heat discoloration. Heat damage cannot be machined out of steels. When there's heat damage, temper in the steel is lost and they should not be reused. And because steels are cheap, there's no excuse for not replacing them.

14 Clean Parts

Use a heated chemical washer to clean all transmission parts.

Professional Mechanic Tip

15 Remove Clutches With C-Clamps

Like the C4's clutches, you can use C-clamps for C-clip and spring removal instead of this spring compression tool found in most transmission shops.

C4 Cruise-O-Matic/Select-Shift

Ford's C4 Cruise-O-Matic and Select-Shift 3-speed automatic transmission is a light-duty unit intended for six-cylinder and small-block V-8 engines. Versions of the C4 have also been produced for big-blocks. The C4 was produced from 1964 through 1981 at Ford's Sharonville, Ohio, transmission plant. An interesting footnote is the C4's status as the first automatic transmission ever designed and actually produced by Ford Motor Company. It is one of the easiest automatic transmissions to rebuild and modify because it is so simple and there are a lot of seasoned racers and builders who know a lot about this transmission and how to get it to perform.

Like GM's Powerglide and Turbo-Hydramatic 350, and Chrysler's 904 Torqueflite, Ford's C4 is a simple light-duty transmission you can knock down and rebuild in your home workshop. There are different approaches as to how to build a C4 and what to put in it. Seems everyone has a different opinion about how to build one for street and competition.

The C4 is an extremely rugged 3-speed automatic transmission capable of withstanding tremendous amounts of power. Even with the best factory parts, the C4 is a hardy box that will serve you for years and thousands of miles with regular fluid and filter changes. As a drag racing automatic, there are a lot of drag-tested components out there designed to enable a C4 to withstand 600 to 1,000 hp on the quarter-mile, making the C4 competitive with GM's legendary 2-speed Powerglide. Savvy transmission guys in the industry have told me they've seen a C4 take as much as 1,200 hp, though I wouldn't suggest pushing a C4 that far. Anything beyond 600 hp presents risk, but that's racing. Street and weekend-racing C4s can easily handle 250 to 600 hp without breaking a sweat.

During the C4's long production life, it underwent several changes you should know about before getting started. From 1964 to 1969, the C4 had a .788-inch, input shaft 24/24-spline, which can take up to 400 hp. In 1970, Ford went to a larger .839-inch, 26/26-spline—one year only. It's very hard to find a 26/26-spline shaft if you're building a 1970 core and need an input shaft. The following year, 1971, Ford changed the input shaft again to a split 26/24-spline type, which means the torque converter end gets a 26-spline and the forward clutch hub gets a 24-spline.

The 26/26 and 26/24 shafts take up to 600 hp, though you should go with a hardened, aftermarket 26/26-spline shaft from TCI Automotive or Performance Automatic if you're going to push it beyond 450 hp. Performance Automatic's hardened input shafts, as one example, are forged from 4340 steel and are engineered for up to 185,000-psi tensile strength, which matters greatly if you're going racing.

Knowing what you have is important when you're looking for C4 cores and parts. Ideally, you can find a complete 1971-or-newer C4 transmission core that already has these changes because very little changed through the end of C4 production in 1981. This means there is an abundance of C4 cores available from 10 years of production so you can build the kind of transmission you need for street or strip.

There were two C4 shift patterns employed during its production life. From 1964 to 1966 came the Green Dot Dual-Range shift pattern transmission. In normal driving, the shifter should be positioned on the large green dot at the detent where the transmission does a routine 1-2-3 upshift. If you want to start out in second gear with the Dual-Range C4, you place the shifter at the small dot (off detent, next to neutral) where the transmission starts out in second gear and upshifts into final drive. The small-dot position was a feature designed to help motorists get started on snow and ice. You don't want the Green Dot valve body for a good street/strip C4 unless you're concerned about originality in a 1964–1966 Ford or Mercury restoration.

Beginning in 1967, Ford went to a more conventional P-R-N-D-2-1 shift pattern for the C4. It was known as Select-Shift, which meant a redesigned valve body that was used through 1969. This is the valve body you want for your 1964–1966 C4 street/strip transmission because it drops right in without modifications. You have to change your shifter to a P-R-N-D-2-1 with the change to a 1967–1969 valve body. If you start out in second gear with this 1967-on pattern, there is no upshift out of second, which makes the 1967-on Select-Shift shift function differently than the earlier 1964–1966 Green Dot, which does upshift.

For 1971-on, there was yet another revised valve body (eight-bolt versus nine-bolt) along with a corresponding change in the transmission case. This means the nine-bolt 1971-on valve body works with the 1964–1970 eight-bolt C4 case with the added bolt and a locknut. However, the eight-bolt valve body does not work with the nine-bolt 1971-on case. My best advice is to go with a 1971-on transmission case and valve body rather than try to mix the two.

C4 Identification

C4 transmissions were originally fitted with factory identification tags located at the intermediate servo cover. Because so many of these tags have been discarded by transmission builders through the years, identification often has to be determined in other ways. And in some instances, it really doesn't matter because castings and parts speak for themselves with numbers of their own. What matters most is the case, bellhousing, tailshaft, tailshaft housing, and key internal parts like the forward clutch and input shaft. It must all work together smoothly and reliably, which depends on you using compatible parts. Identification can be determined through Ford casting numbers, date codes, and visual identification of these components.

Case-Fill or Pan-Fill

Although there were a variety of C4 castings produced between 1964 and 1981, there are but two basic types you can expect to find: stepped and blended-bell cases with a wide variety of tailshaft housings

Core Choices

When you're searching for a rebuildable C4 core, aim for 1971 and newer as a base transmission core, though most cores you're going to find are comprised of parts from across different model years from rebuilds through the years. You want a C4 core with the 26/24-spline input shaft and forward clutch plus the nine-bolt valve-body main case. This approach keeps your project simple and easy to manage. ■

and bellhousings. The stepped bell, also known as case-fill, has a seven-bolt (bellhousing-to-case) bellhousing that bolts to the front pump and is designed for compact and intermediate applications, though it can be installed in full-size cars and trucks. The case-fill has a dipstick tube provision in the left-hand side of the case, hence the term case-fill.

The blended-bell case is known as a pan-fill because the dipstick tube goes into the pan instead of the case. This type is for heavy-duty applications in full-size cars and trucks. The pan-fill/blended bellhousing bolts to the main case instead of the front pump and has a larger five-bolt bell-to-case pattern for less noise, vibration, and harshness, along with better load distribution.

The pan-fill C4 does not fit compact and intermediate Fords and Mercurys due to dipstick tube clearance issues with headers and frame rails. If you find a pan-fill C4 for your Mustang, Fairlane, Torino, Falcon, or Comet, it won't fit. Instead, find a stepped bell/case-fill core because fitment is nearly impossible and always frustrating with the pan-fill C4. Seasoned transmission builders view the pan-fill C4 as the strongest case available. However, fitment is tricky to impossible in compacts and intermediates, which leaves you with case-fill only.

Another main case option available to you is the 1982–1986 C5, which is compatible with the C4 in terms of key components that interchange. The C5 case has better hydraulic/lubrication circuits along with larger cooler-line fittings. When I ask transmission builders what they think of C4 versus C5 main cases, most take the C5 over the C4 when possible due to improved hydraulic circuitry and cooling. However, most also say there's little difference between the two unless you're building an all-out drag racing transmission.

The C5 was produced in both stepped-bell and blended-bell configurations as well as 157- and 164-tooth flexplates. You may even use the stepped-case C5 with a 148-tooth flexplate for Mustang II and Pinto.

Bolt Pattern

Another item to watch for is five- versus six-bolt bellhousing-to-engine-block bolt patterns with C4 transmissions. Although finding an early-1964-only, five-bolt C4 bellhousing is a long shot these days, they're out there and are quite valuable because so few remain. The problem is, a five-bolt bell does not fit a six-bolt block. Unless you have a 1962–1964 221-, 260-, or 289-ci engine with a five-bolt block, you need the improved six-bolt C4 bellhousing.

Another bellhousing fact that can catch you off guard is the wider C5 bellhousing used only on the 1982–1986 C5. Because the C5 has a locking-clutch torque converter for improved efficiency, the converter is wider and, therefore, needs a wider six-bolt bellhousing. If you use the C5 bellhousing on your C4 and try to fit it into a classic Mustang, you're in for a rude awakening because it does not clear the tunnel. It does not match up with your flexplate, front pump, or starter either. Look for E2, E3, E4, E5, and E6 casting numbers on this bellhousing. It is just as easy to use a C4 bellhousing on a C5 case because you're not going to be using a locking torque converter with C4 components.

The C4 case-fill is the most common C4 bell out there. This is the case you want for compacts and intermediates like Mustang, Fairlane, Falcon, Torino, and Comet because it clears frame rails and headers.

Performance Tip

The case-fill looks like this on the right-hand side, with provision for a dipstick tube. This particular C4 has a C4OP-D027-B B intermediate servo, which indicates a six-cylinder application. For improved performance, you want either the H or R servo and piston designed for full-sized Fords, or the C servo for 289 High-Performance V-8s. Because used C servos are hard to find, go with the larger-capacity H servo. Reproduction C servos are available from Scott Drake Reproductions and Mustangs Plus.

The case-fill seven-bolt bellhousing bolts directly to the front pump.

Flexplate Match

Another important consideration is the size of flexplate and bellhousing you intend to use. The C4 was manufactured with a variety of bellhousing and flexplate configurations—157- and 164-tooth are the most common. You probably see more 157-tooths than anything because the C4 was installed in more compacts and intermediates than any other Ford transmission.

The 164-tooth flexplate and bellhousing were more common to full-size cars and trucks, though they were fitted to 289 High Performance V-8 applications in compacts and intermediates. The main consideration is to make sure you have a bellhousing and flexplate that match. Where building a C4 becomes tricky is for the 1974–1978 Mustang II, 1971–1980 Pinto or Bobcat, and any Mercury Capri before 1979. Because the transmission tunnels on these particular vehicles were small, Ford developed a smaller 148-tooth bellhousing/flexplate combination. Most of you know Pinto, Bobcat, and Capri were never factory fitted with a V-8. However, a few of us like V-8s in these classic subcompact Fords and Mercs. Whatever subcompact platform you choose for your project,

The blended bell pan-fill C4 with 164-tooth flexplate was designed for heavy-duty applications like large cars and trucks and was employed right from the start in 1964. The blended-bell C4 does not fit a compact or intermediate unit-body Ford or Mercury due to frame rail and header clearance issues.

This is the blended-bell pan-fill case for full-size cars and trucks. It presents dipstick tube clearance problems in subcompacts, compacts, and intermediates. This blended-bell C4 has a C4OP-7D027-A A intermediate servo for V-8 engines. It is the most common intermediate servo you can find. But the A servo isn't the one you want. Aim for an H or R servo cover and piston.

you're going to need the smaller 148-tooth bellhousing and flexplate.

Where it becomes frustrating is trying finding a 148-tooth bellhousing/flexplate combo anywhere. There may be a few laying around, but most are gone, victims of clunker laws and crusher programs. Salvage yards remain among the few sources for these rare pieces. If all else fails, one of the best sources is Transtar for new parts. For used parts like bellhousings, there are Ford specialty parts dealers around the country all too happy to help you.

C4 Operation

In order to best understand how to build a C4, you must first know how it operates. The C4 transmission employs a conventional fluid coupling known as a torque converter. The torque converter connects the engine's crankshaft with the transmission's input shaft via a fluid coupling, and it multiplies the engine's torque under acceleration with a process as time-proven as a good-old-fashioned water wheel. With the shifter in neutral, the torque converter, input shaft, and forward clutch whirl around at the engine's speed. Hydraulic fluid under pressure continues to circulate through the front pump, the torque converter, and return to the sump. Unless you place the shifter in one of three ranges, this is how the transmission operates with the engine running.

"Drive" Range

When you place the C4's selector in Drive, fluid under pressure is routed to the forward clutch beginning at the manual shift valve. From the manual shift valve on, it gets complicated.

Engine power travels through the forward clutch to the forward ring gear hub and ring gear. When power is applied, it flows to the forward ring gear, which drives the front planet carrier, which is splined to the output shaft. The front planet pinions then rotate clockwise and the sun gear rotates counterclockwise. As the sun gear rotates counterclockwise, you get clockwise rotation of the reverse planet carrier.

First Gear

Because the reverse planet is held stationary by the one-way roller clutch and reverse band, the output shaft ring gear then rotates around the reverse planet pinions. Output shaft ring gear rotation is then transferred to the output shaft ring gear hub.

This rotation then causes the forward planet to rotate clockwise, only slower than the ring gear for a 2.46:1 ratio (2.46 revolutions to 1 revolution of the output shaft). If the selector is placed in Low, the low-reverse band is applied, which then makes engine braking under deceleration possible.

Second Gear

When there's sufficient speed, the C4 upshifts into second gear. Second gear also happens when you manually place the selector in second or Drive 2 (DR2). When this happens, there's forward clutch engagement (already happening in first gear) and intermediate band application (second gear only).

With this turn of events, power flows through the forward clutch to the front planetary-ring gear hub, which is tied to the output shaft. Power stops at the front-unit ring gear. The ring gear then rotates around the front planet gears, which are tied to the front planet carrier, which is tied to the output shaft. Front planet gears rotate around the stationary sun gear, which is connected to the input shell. The sun gear is held still by the intermediate band, which is locked around the reverse-high clutch drum. The reverse-high clutch drum is connected to the input shell.

TECH TIP

Core Options

When you're shopping for a C4 core for your Ford subcompact, compact, or intermediate, choose the stepped-bell/case-fill unit. The blended-bell/pan-fill C4 does not fit a subcompact, compact, or intermediate. ■

At this point, power is flowing through the front planet carrier to the output shaft for a gear reduction of 1.46:1 (1.46 turns of the input shaft for every 1 rotation of the output shaft).

High Gear

When it is time for upshift into direct or final drive, both the forward clutch and reverse-high (direct) clutch are applied. As both clutches are applied, the entire geartrain is turning in the same direction at the same speed meaning input and output shafts are turning at the same speed in the same direction. There is no gear activity in direct or final drive for a 1.00:1 ratio. One full rotation of the input shaft equals one full rotation of the output shaft.

Blended-bell pan-fill C4 transmission bellhousings bolt to the main case (instead of to the front pump) with seven bolts and a larger bolt spread to reduce noise, vibration, and harshness while providing strength.

Reverse Gear

In reverse gear, you apply the reverse-high clutch pack and low-reverse band. Power flows to the input shaft through the forward clutch to reverse-high clutches and drum to the input shell, which is tied to the sun gear. The sun gear is driven by the input shell, which drives the reverse planet gears. The reverse planet pinion gears drive the reverse ring gear in the opposite direction.

In other words, the input shaft turns one way and the output shaft is driven in the opposite direction.

The low-reverse drum is locked by the low-reverse band, which keeps the rear planet carrier from turning.

On the left is a blended bell with the five-bolt bell-to-transmission bolt pattern. On the right is the more common stepped bell with a seven-bolt bell-to-pump bolt pattern. Both are for 157-tooth flexplate.

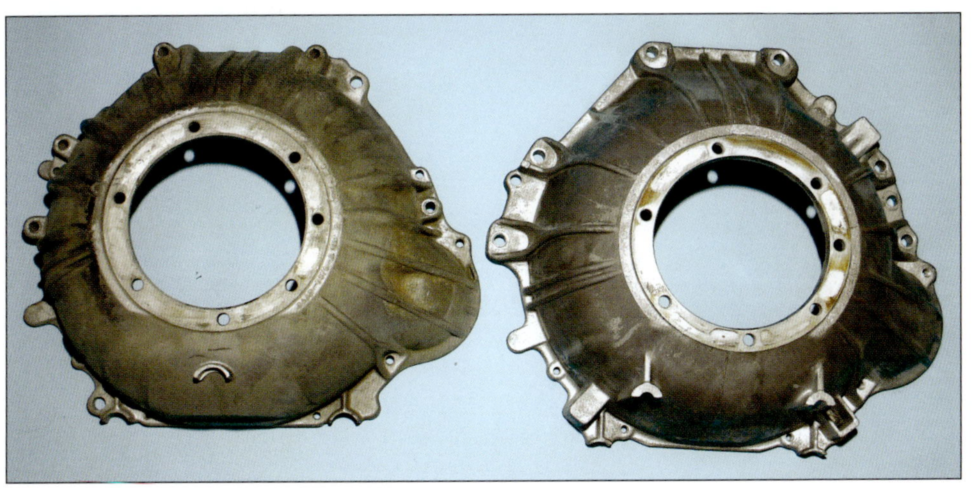

Another issue to watch out for is five- versus six-bolt bellhousings. Although five-bolt bells are scarce these days, they do exist and they can stump you if you're not paying attention. The smaller five-bolt bell (left) is for five-bolt small-block V-8s prior to August 1964. The larger six-bolt pattern (right) was conceived for the 1965 model year to reduce noise, vibration, and harshness while also providing improved strength.

Power flows through the reverse planet pinions to the ring gear, which drives the output shaft. This gives you a 2.20:1 ratio (2.20 turns of the input shaft for every 1 rotation of the output shaft).

Building a Better C4

Because you want the most reliable performance possible from your C4 build, you'll want to choose the best components available. Leon's Transmission has chosen B&M Racing & Performance for this C4 build. And because there's no real magic going on here, knowing what to use is a matter of doing what the experts who build high-performance C4 transmissions do. Because most of you reading this book are building street and weekend race transmissions, you're going to want reliable feedback.

Experts and racers generally believe the blended pan-fill C4 case is the strongest C4 case; however, it faces space limitations with subcompact, compact, and intermediate Fords. This means you are limited to the stepped-bell, case-fill C4 for your less-than-large Ford. The news isn't all bad however.

If originality and matching numbers aren't important to you, go with a 1971-on C4 core with a 26/24-spline input shaft and forward clutch, which is the easiest C4 generation to build in terms of parts and availability. You may also go with the 1982–1986 C5 stepped-bell, case-fill main case and C4 bellhousing to achieve the most reliable C4 possible. All you have to do is fill it with C4 internals and make the necessary modifications, which are minimal.

When it comes to case-fill versus pan-fill, the C4's dimensions speak for themselves. The blended bell, pan-fill C4 does not fit your compact or intermediate without serious compromise. But with the right combination of parts, a stepped-bell, case-fill C4 offers convenience and is capable of staying together at more than 600 hp.

After you've selected the right C4 main case and bellhousing for your project, turn to what goes inside. Unless you are committing your C4 transmission to full-time drag racing, you're not going to need a manual valve body or a trans brake. The main focus of this book is street and occasional drag racing.

If you have a 250- to 350-hp small-block or a six, a stock C4 with the standard complement of V-8 clutches (sixes have fewer) and a shift improvement kit delivers the kind of performance you need for the street. Reliability comes from a firm shift instead of a soft shift, where slippage and friction burning occur. You don't want a harsh shift unless you're going racing. The key to reliability is the type and number of clutch frictions and steel thicknesses you're going to use, along with setting proper clearances.

Intermediate Servo

The C4 has been factory fitted with a variety of intermediate-band servos for an even wider range of applications. The most popular intermediate-band servos are the C, H, and R, which have the largest piston and bore size for maximum band pressure. The most common is the A servo found on most small-block C4s. Six-cylinder units have the B servo, which isn't recommended.

Though racers and builders have a real obsession with the C, H, and R servos, in truth there's not much difference in performance from one

Another C4 issue to watch out for is transmission ventilation. Early C4 transmissions were vented with a steel tube from the center case on the left-hand side. This means the tailshaft housing is not vented.

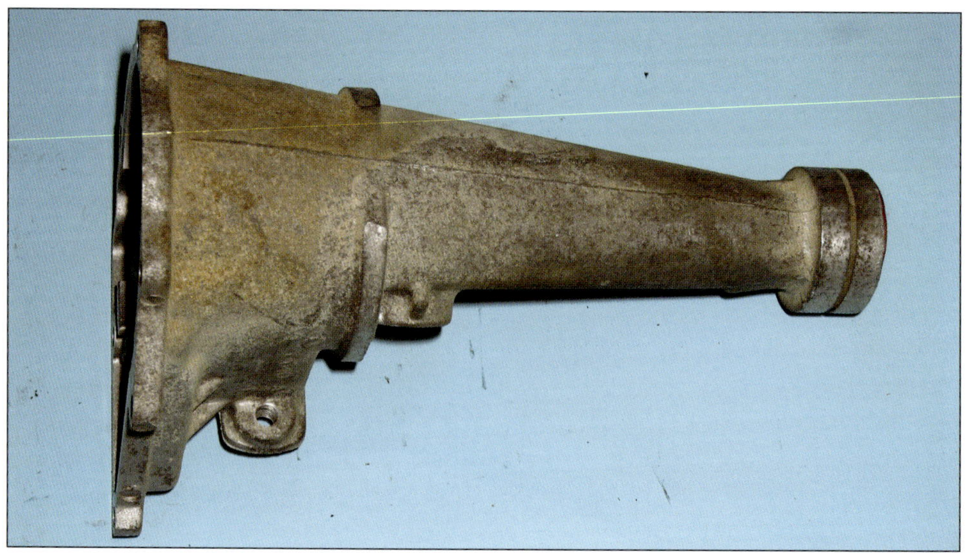

servo size to the next. Transmission professionals I've worked with have proven the H and R servos provide the best intermediate-band apply pressure, which is based on having the largest cover and piston size.

An added twist to servo selection is how quickly the servo releases the intermediate band during 2-3 upshifts. You want the quickest release possible at wide-open throttle without RPM flare. This is why the R servo is more favorable than the H, if you can find one.

You want the most piston/bore surface area possible for maximum hold. However, what also matters a lot with the intermediate band is how wide the friction area is coupled with the type of friction material. Drag racers prefer Kevlar and Blue intermediate bands because they hold and release better, and last much longer in tough racing competition. Kevlar makes very little sense for the street considering the cost involved and minimal benefit gained.

Because the low-reverse band doesn't get the kind of workout the intermediate band does, I'm not concerned with making changes here because it has absolutely no real effect on performance.

The C5 transmission case looks identical to the C4 case and can be used for a C4 buildup. Expect to see E2, E3, E4, E5, and E6 Ford casting numbers on C5 cases. The C5 differs with improved hydraulic circuitry and more generous oil flow. The C5 also differs from the C4 in the transmission cooler-line fitting sized at 1/4 inch. C4 band adjustments are coarse-thread and C5s are fine-thread. The C5 also has a deeper pan than the C4 along with a corresponding pickup.

Be careful with this one—it is a wider C5 locking torque converter bellhousing, which is wider to accommodate a wider converter. And if you think you can bolt a C5 into your classic Mustang, Falcon, or Comet, keep in mind this bellhousing does not clear the transmission tunnel.

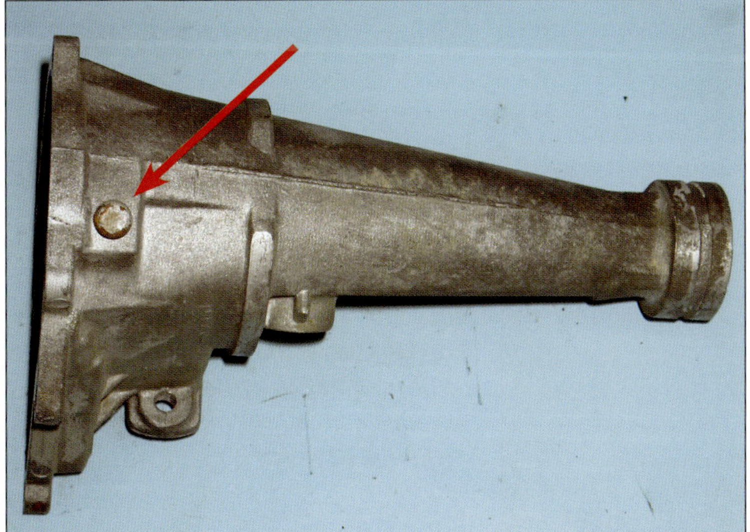

C4s without the center case steel tube vent are vented in the tailshaft housing with this little "mushroom" vent.

Tailshaft length and housing are important considerations. This short tailshaft housing is for an application where it is supported from above instead of a crossmember underneath.

Front Pump

In all the transmission builds I've seen through the years, I've never witnessed modifications to the front pump. They just seem to keep going and going. The main concern is cavity and gear condition, along with proper operation of the pressure-relief and drain-back valves.

Check fluid passages for debris. Always replace the bushing and seal. The front pump is just that—a positive-displacement hydraulic pump that provides control pressure and lubrication. This means it doesn't require much attention aside from inspection of gears, cavity, and valves, plus replacement of bushing and seal. Unless a pump ingests foreign matter or becomes cavitated during operation, there's little chance of damage or extreme wear.

Valve Body

C4 valve bodies boil down to four basic types: 1964–1966 Dual-Range, 1967–1970 eight-bolt, 1971-on nine-bolt, and 1971–1980 Pinto/Bobcat (also nine-bolt). There are other minor differences too numerous to mention. What you're concerned with most is basic valve body types.

Trans-Go, B&M, and TCI Automotive offer shift improvement kits for the C4 that not only firm up shifts, they also add longevity because they reduce and eliminate clutch and band slippage. They enable the valve body to apply more pressure to band servos and clutches. It is also the timing of clutch and band application and release that makes these kits so effective.

Throttle Valve Operation

Working with the valve body is the C4's throttle valve, also known as the vacuum modulator. Your C4's throttle valve senses engine load and the need for torque by getting its signal from intake manifold vacuum, which acts on the throttle valve's diaphragm, spring, and rod. When vacuum drops (open or wide-open throttle), the throttle valve increases control pressure through the valve body and contributes to the delay of upshift at wide-open throttle. When intake manifold vacuum is high (closed throttle), the throttle valve

C4 flexplates exist in four basic sizes. From left to right are 164-tooth, 157-tooth, and six-cylinder. Starter drive ring gear is on the torque converter in 144/170/200-ci six-cylinder applications. Not shown is the 148-tooth flexplate for Mustang II, Pinto, and Capri applications with smaller bellhousing.

There are input shaft/forward clutch changes in the C4 you should be aware of. From 1964 to 1969, the C4 had a .788-inch 24/24-spline input shaft and forward clutch hub (left). For 1970 only, there was a larger .839-inch 26/26-spline shaft and forward clutch hub (not shown). From 1971-on, Ford went to an .839-inch shaft with a 26/24-spline configuration (right). If you're going racing, go with a hardened aftermarket 26/26 shaft from Performance Automatic.

Bellhousings

Watch out for those early five-bolt bellhousings for 221, 260, and 289 V-8s. Although it is unlikely you will find one, a five-bolt bell will mess up your afternoon when it's time to install the transmission.

One other item to watch out for is the Pinto, Mustang II, and Capri small six-bolt bell and 148-tooth flexplate. Unless you're building a C4 for one of these car lines, you're going to want the 157- or 164-tooth flexplate and bell. ■

helps the valve body lower control pressure and prompt a more immediate upshift.

From a performance standpoint, there's not much you need to do to the throttle valve, aside from choosing one you can adjust and tweaking its adjustment to suit proper transmission function. Remember, the throttle valve is important to transmission control pressure, which means being careful with how you make adjustments. Poor vacuum valve adjustment can lead to transmission failure because control pressure is critical here. Hindrances to control pressure can cause slippage, heat, and wear. This is why this valve is so important to proper function.

To adjust the throttle valve properly, you've have to know the control pressure, which involves a pressure gauge at the control pressure port just above the manual shift and neutral safety switch. Transmission sump temperature should be at hot idle. Cold transmission fluid gives you an erroneous pressure reading on the high side because the fluid is more dense. The gauge's pressure range needs to be 0 to 400 psi.

First, you must establish proper throttle valve function. Does the throttle valve work properly? Is there vacuum leakage? This is checked with the throttle valve removed using a vacuum pump or an intake manifold vacuum from an operating engine at idle. At 18 to 22 inches of vacuum, you are able to tell if there's leakage by a hissing sound or the absence of diaphragm/rod movement. There should also be diaphragm and rod movement at 18 to 22 inches of vacuum. At higher elevations, your engine may struggle to maintain 18 inches at idle.

Valve body identification can get tricky, but not if you know what to look for. Confusion abounds when mass rebuilders get pieces mixed up. Here are early C4 valve bodies—on the left is a Dual-Range valve body for Green Dot transmissions, which no one but a restorer wants because they don't make sense for performance applications. On the right is a 1967–1970 valve body with the P-R-N-D-2-1 shift pattern, which fits only 1964–1970 C4 cases with the eight-bolt pattern. Beginning in 1971, Ford changed both the case and valve body to an nine-bolt pattern making them incompatible with 1964–1970 units.

With the throttle valve installed, transmission in neutral/park, and engine at idle, you should have a minimum of 18 inches of manifold vacuum. Anything lower than 18 inches of vacuum indicates engine health problems that must be corrected or you have a really lumpy camshaft profile. If you have at least 18 inches of vacuum and the selector is in any forward gear, you should see 55 to 62 pounds of line pressure (55 to 100 pounds in reverse).

With the throttle open at 10 inches of vacuum and the transmission in any forward gear, you should see 96 to 105 pounds of line pressure. At 3 inches of vacuum, expect to see 138 to 148 pounds of line pressure in any forward gear. In reverse at 3 inches, expect to see 215 to 217 pounds of line pressure.

Whenever line pressure isn't within these parameters, throttle valve adjustment must be done with great care. When line pressure is too high, shifts become harsh. If line pressure is too low, shifts become soft (slippage) and that's when there is damage with burned clutches and bands. To increase control pressure, throttle valve adjustment needs to be clockwise. To reduce line pressure, counterclockwise. Ford says one full turn either way adjusts line pressure 2 to 3 psi. Check line pressure at idle, then, at 10 and 3 inches of manifold vacuum to determine if any further adjustment needs to be made.

Never base your adjustments on shift feel alone because that isn't wholly what line pressure is about. Always use a pressure gauge and follow Ford's guidelines to the letter.

Building the C4

Although automatic transmissions can be very intimidating, they really are engineering marvels that have only gotten better with time and improved engineering. And though they've become more complex, their basic concept hasn't changed much in a half century. The C4 remains one of the simplest automatic transmissions ever produced, which means you can build this transmission with confidence if you follow instructions and take your time.

Because manufacturers, such as B&M Racing & Performance, have developed kits and parts for Ford automatic transmissions, it has become easier to source what you need to perform a rebuild in your home workshop. And, if B&M doesn't have it, Transtars are available at a host of local transmission parts suppliers from coast to coast.

When sourcing parts for your C4 build, never cut corners in your efforts to save a few bucks because it isn't worth sacrificing durability. You get what you pay for. Do it on the cheap, and you have to go back inside sooner. In few places is this more true than clutches and bands.

You want the best friction materials possible. This doesn't mean going with racing clutches and bands in your C4 because, although they work quite well short-term, they don't last long-term. The best street frictions are those red factory original pieces.

Your performance rebuilding plan should include the 1967–1969 or 1970-on valve body (depending on which case you have) along with a proven shift-improvement kit.

This is the stepped-bell case-fill C4 with valve body and servo struts removed.

C4 Function Chart

Selector	Gear	Ratio	Forward Clutch	Reverse-High Clutch	Intermediate Clutch	Low-Reverse Band	One-Way Clutch
P or N	Neutral	None	Off	Off	Off	Off	Off
L	Low	2.46:1	On	Off	Off	On	Hold
D1	Low	2.46:1	On	Off	Off	Off	Hold
D1 or D2	Second	1.46:1	On	Off	On	Off	Over Running
D1 or D2	High	1.00:1	On	On	Off	Off	Over Running
R	Reverse	2.20:1	Off	On	Off	On	N/A

The intermediate servo and band are important to upshift quality and transfer of power. Although there's a lot of hype over the C 289 High Performance intermediate servo, the H and R servos for full-size cars and trucks are also an excellent choice because they're larger than the C servo. The O servo shown here on the right is an E2TP C5 truck intermediate band servo. The main thing you want to pay attention to is cover and piston size along with Ford part number. The aftermarket also provides you with larger billet-aluminum intermediate servos for improved performance.

Kevlar bands work quite well for all-out drag racing but make no sense on the street. And their cost doesn't justify their use on the street either.

Your C4's greatest asset isn't always the parts, but the assembly technique practiced during build-up. It is those details missed that come back to haunt later. Again, remember what you observed during the disassembly. The cause of transmission failure can be an important detail missed during assembly. Prior to assembly, closely re-inspect all parts for unusual wear patterns or damage.

One thing you want to be prepared for is small changes that happened in C4 production, such as low-reverse clutch-return spring design. Early in production, Ford used 10 low-reverse clutch-return springs. Later, in the 1970s, Ford went to one large return spring to reduce costs and improve clutch performance.

The reverse-high clutch assembly does not have to involve special tools. You may use four C-clamps to compress the spring and install the C-clip. When you do this, take care to protect your eyes and face.

During geartrain assembly, it is easy to get lost in the details. Thrust washers, which take up endplay and give components thrust support, are easy to overlook. Instead of thrust washers, late-model automatic transmissions like the AOD, AODE, and 4R70W have Torrington (roller) bearings, which reduce power-robbing internal friction throughout the geartrain.

If you want to reduce internal friction your C4 can be fitted with Torrington bearings. However, Torrington bearings, and modifications designed to accommodate them, do not come cheap. But if you're going racing, they make your C4 more efficient and add power.

More for Your C4

Although this book is slanted toward the street and occasional strip enthusiast, it's always nice to know there's help if you want a little something more from your C4 build. Because the C4 is an extremely rugged box that takes a lot of punishment, you can run a lot of power through it, even using factory pieces such as clutches and bands.

The best pieces are factory C4 components that came along after 1970. And you just can't beat what the aftermarket has come up with since 1970. Ford's C4 enjoys the same notoriety with racers as GM's legendary Powerglide, thanks to its lightweight, back-yard simple design. And there are plenty of them out there.

This is the low-reverse clutch servo cover and piston.

C4 Line Pressure Quick Reference

Intake Manifold Vacuum at Idle (inches)	*Line Pressure (psi)*
17	55–62
16	55–68
15	55–74
14	55–80
13	55–87
12	55–93
11	55–99

C4 Servo Covers and Pistons

When it comes to C4 intermediate-servo covers and pistons, there are more combinations than I could ever get into in one book. So I'm focusing on basic types and sizes without getting into every servo cover and piston part number Ford has ever issued. These are parts you are most likely to encounter. But be advised, there are many others out there.

Intermediate Servo Cover Identification

Identification	*Part Number*	*I.D. (inches)*
A: C4OP-7D027-A	C4AZ-7D027-F	2⅞
A: C6AP-7D027-C	N/A	N/A
B: C6AP-7D027-B	C4OZ-7D027-E	2¾
C (289 Hi-Po)	C5AZ-7D027-B	2⅞
H	C8ZZ-7D027-A	2$\frac{15}{16}$
K	D2DZ-7D027-A	2⅜
R	D0AZ-7D027-A	2⅞
W	C5DZ-7D027-B	2$\frac{17}{32}$
Y	C5ZZ-7D027-B	2¼
Z: C6DP-7D027-D	C4AZ-7D027-G	2$\frac{17}{32}$

Intermediate Servo Piston Identification

Identification	*Part Number*	*Seal O.D. (inches)*	*Inner Seal O.D. (inches)*
A	C5AZ-7D022-B	2⅞	3$\frac{13}{16}$
B	C5AZ-7D022-A	2⅞	3$\frac{13}{16}$
C	C5AZ-7D022-C	2⅞	3$\frac{13}{16}$
H	N/A	N/A	N/A
K	D2DZ-7D022-A	2$\frac{17}{32}$	3⅜
R	D0AZ-7D022-A	2⅞	3$\frac{5}{16}$
W	C5DZ-7D022-A	2$\frac{17}{32}$	3⅜
Y	C5FZ-7D022-A	2¼	3¼
Z	C5OZ-7D022-A	2$\frac{17}{32}$	3⅜

Note: These are the most common combinations found in the Ford Master Parts Catalog. There may be some unaccounted for. C5 intermediate servo covers and pistons are not shown here.

The C4's front pump is a positive-displacement gear pump that provides both hydraulic pressure and lubrication. The pump's drive gear is driven by flats on the torque converter. The drive gear hauls the driven gear around, drawing in fluid from the transmission's sump. It delivers fluid to the main pressure control system. The control pressure regulator keeps pressure safe by venting fluid back to the sump, which is a normal part of operation.

Most transmission professionals like to call this valve the vacuum "modulator." Ford calls it the throttle valve. This valve is affected or modulated by throttle position and engine load, which directly relates to intake manifold vacuum. When vacuum is low (throttle open), this valve signals shift valves to delay upshift and the main pressure regulator to increase control pressure while the vehicle is under acceleration. When vacuum is high (throttle closed), control pressure is reduced and shifts occur sooner. When spring pressure exceeds vacuum pressure, you increase control pressure. When vacuum exceeds spring pressure, you reduce control pressure. The throttle valve is adjustable in order to control line pressure.

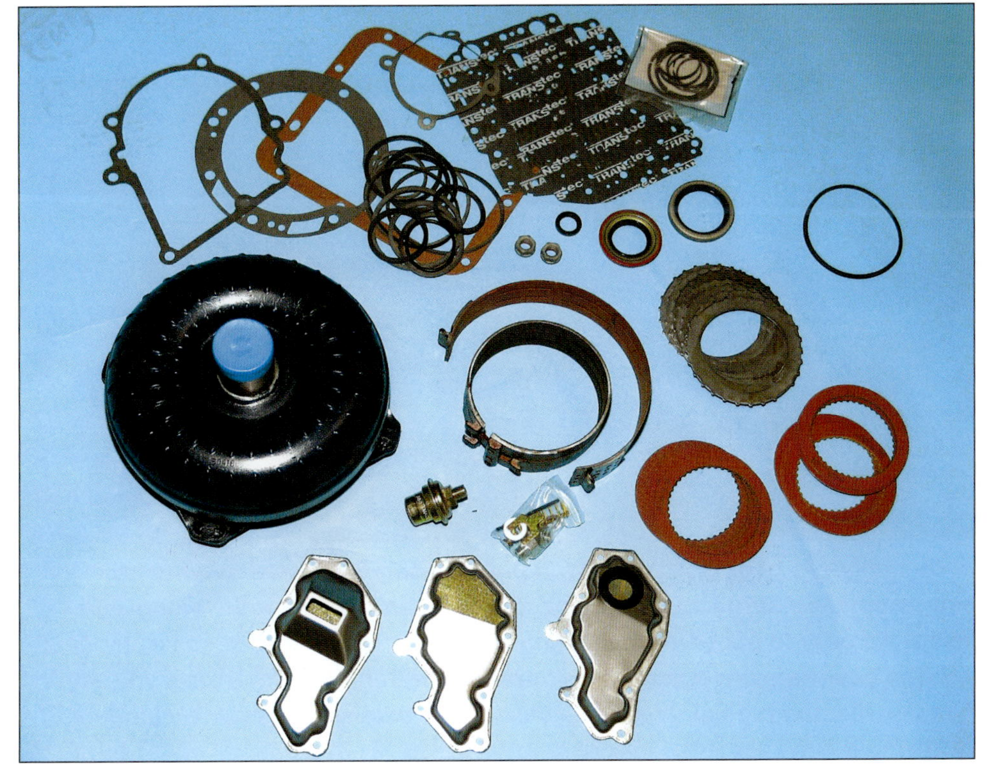

Leon's Transmission is going to build a C4 using a complete B&M Racing & Performance Transkit (PN 50231) for the 1970–1982 C4. The B&M Transkit consists of clutches including steels, bands, gaskets, seals, throttle valve, and filters. Hard parts, such as planetaries, shafts, clutch drums, and the like, can be sourced from Transtar.

C4 Line Pressure Guide

Engine Vacuum (inches)	Throttle Status	Shifter Selection	Control Pressure (psi)
18 and Higher	Closed/Idle	Forward Ranges and Neutral Only	55–62
18 and Higher	Closed/Idle	Reverse Only	55–100
In Transition at 17 and Lower	Opening	Forward Ranges Only	Should show a rise in line pressure
10	Open	Forward Ranges Only	96–105
3	Open	Forward Ranges Only	138–148
3	Open	Reverse Only	215–227

Be choosy about the type of aftermarket transmission parts you use. Buy cheap transmission parts and you can expect shorter transmission life. Always go with the best materials available for long life. On the street, you want longevity. For racing, longevity isn't as important as performance.

Reassembly

1 Begin Assembly

Assembly on this 1974 Ford C4 Select-Shift transmission begins with the front pump assembly. Install bearing race and front seal first, with generous lubrication on both. Be sure you have inspected drive and driven gears along with the case, checking for abnormal wear patterns. It's easy to get gear installation backward, where torque converter flats won't engage driven gear flats. Make sure driven gear flats are open to the torque-converter side. Otherwise, you won't be able to seat the torque converter, which drives the pump. Pack pump cavities with transmission assembly lube for generous lubrication on startup.

Critical Inspection

2 Inspect Sealing Rings

Pump stator support sealing rings and contact surfaces should be closely examined for wear, scoring, and proper fit. This means both stator support and forward clutch contact surfaces. Some builders use Teflon sealing rings with great success. But Teflon is more appropriate for racing applications, though there are bound to be arguments on this one. Much depends on how you intend to use your C4 transmission. Teflon seals are easier than iron rings on the hard parts, but iron rings tend to last longer because there's not as much difference in hardness.

Setting Geartrain Endplay

Seeking proper geartrain endplay? C4 transmission geartrain endplay is the sum total of all your transmission's moving internal parts, which is adjusted at washers number-1 and -2 at the front pump. Endplay should be .008 to .042 inch, which is what Ford calls for. To achieve endplay within factory limits, go with the following information from Ford.

Fiber Washer Number-1

Thickness (inch)	*Identification Color*
.038 to .042	Green
.053 to .057	Tan
.070 to .074	Black
.087 to .091	Yellow
.104 to .108	Blue
.121 to .125	Red
.138 to .142	Purple

Steel/Bronze Washer Number-2

Thickness (inch)	*Identification Number*
.041 to .043	1
.053 to .056	2
.073 to .075	3
.090 to .093	4
.107 to .109	5

The stator support hub gets two thrust washers, number-1 and number-2. These are known as selective-thrust washers and available in various thicknesses to help you get proper geartrain endplay. Washer number-1 is fiber and available in five thicknesses. Washer number-2 is steel backed with a bronze bearing surface and also available in at least five thicknesses—steel backing for strength and bronze for wear. According to Ford, endplay between forward clutch and stator support should be .008 to .042 inch. Be careful how you install washer number-2 because it can block oil passages if installed incorrectly.

Clutch Pack Assembly

1 Begin Forward Clutch Assembly

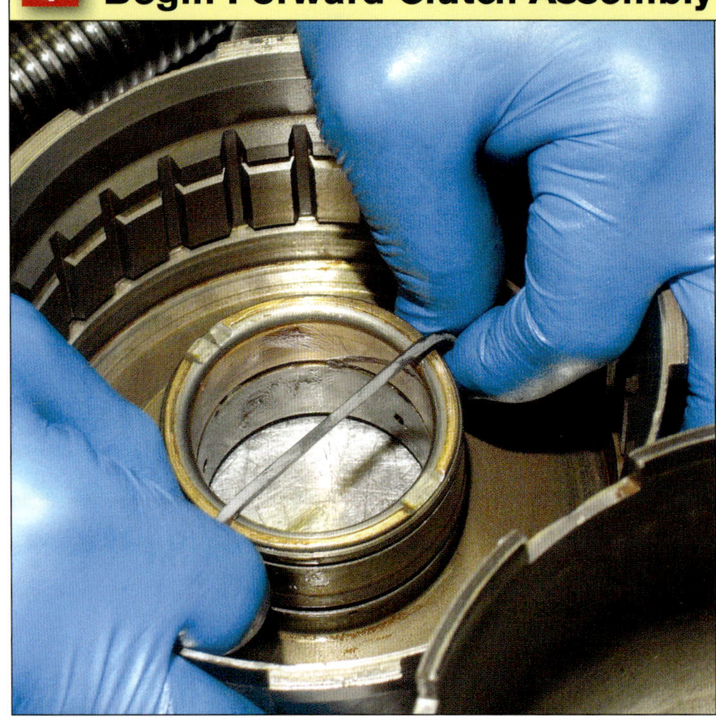

Forward clutch assembly begins with piston seals, which go on the clutch cylinder and clutch piston. Seals get generous amounts of transmission assembly lube. Make sure seals are seated squarely in grooves. A twisted or disfigured seal fails, so make sure seals are seated square as clutch piston is installed.

2 Lubricate Forward Clutch Cylinder

Generously lubricate the forward clutch cylinder with transmission assembly lube to ensure smooth operation on start-up. Dry seals are damaged on start-up. You can never have too much lubrication.

3 Secure Clutch Piston

After forward clutch piston is installed, this snap-ring is next to secure clutch piston.

Critical Inspection

4 Inspect Belleville Spring

This is the Belleville spring, also called a disc spring, which returns the forward clutch piston to rest, releasing clutches when hydraulic pressure is released. This is a good time to inspect the Belleville spring for cracks. Small hairline cracks can cost you time later.

5 Remove Forward Clutch Segments

Forward clutch segments are next, beginning with this steel. There are four or five clutch frictions in a stock C4 V-8 forward clutch. Sixes have four, as a rule. You really want five clutches.

6 Clean Clutch Frictions

Always soak new clutch frictions in transmission fluid before installation. There's no real harm in installing dry clutches, but it makes more sense to get started with saturated frictions.

7 Install Forward Clutch Frictions

Forward clutch frictions and steels installed. Steels are driven by the clutch cylinder (also called a clutch drum). Frictions (clutch plates) drive the ring gears and planetaries. Clutch clearances should be .022 to .042 inch, determined primarily by snap-ring thickness. Selective snap-ring thicknesses are .088, .092, .074 to .078, or .060 to .064 inch. This may vary considerably in the aftermarket. Steel thicknesses also vary, which affects clearances. Be prepared to measure them with a micrometer and then check clearances with a thickness gauge after installation.

8 Install Reverse-High Clutch Piston

Be sure to install a reverse-high clutch piston with new seals and plenty of lubrication. You want things real slippery for a good healthy start-up. Reverse-high clutch return springs do differ in old-versus-newer C4 units. Early units have 10 small clutch-piston return springs, which is what you witnessed during disassembly. This mid-1970s C4 unit has a single large return spring. Be prepared for engineering changes like this in your C4.

Special Tool

9 Install Reverse-High Snap Ring

Use a spring compressor to install a reverse-high clutch-piston snap-ring. You can do this in your home garage with four C-clamps. And this is the only phase of C4 assembly where any kind of special tool or fixture is required.

10 Assemble Reverse-High Clutch

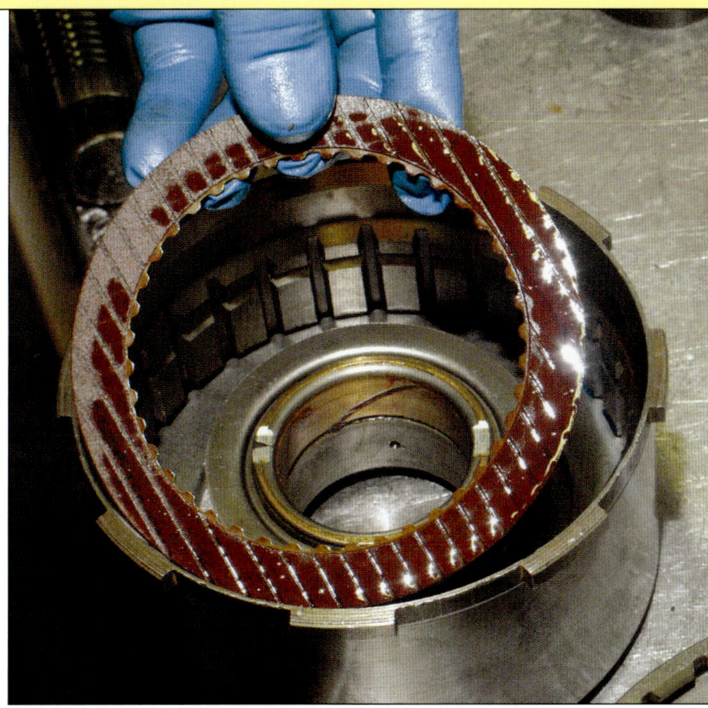

Reverse-high clutch (also called a direct clutch) assembly begins with the first steel for a total of four clutches. Some applications have three frictions. But it is possible to get five clutches in a stock drum/cylinder depending upon both the clutch drum and clutch steel thicknesses. Aftermarket direct clutch drums take as many as seven clutches for improved power transfer without power loss due to slippage. The more friction surface you have going here, the better.

11 Check for Proper Function

Air-check the forward clutch for proper function. Air moves the forward clutch piston, which can be heard and observed as air is applied. Do this with the forward clutch mated to the front pump/stator support assembly.

12 Install Thrust Washer

Prior to installing the ring gear/hub, install the number-3 thrust washer. Use plenty of transmission assembly lube on contact surfaces.

13 Reassemble Planetary/Clutch

Forward planet carrier and ring gear/hub go together along with the number-4 thrust washer on the backside of the carrier.

14 Connect Forward Clutch and Ring Gear

Mate together the forward clutch and ring gear, which takes practice. Spline the forward clutch hub and ring gear into clutch frictions by gently working the ring gear back and forth until fully seated. Note that in this photo the forward planet carrier is not installed in the ring gear. This ring gear drives the forward planet.

15 Connect Planet Carrier to Ring Gear

The forward planet carrier is meshed into the ring gear, which is seated into the forward clutch. (Note: the reverse planet carrier and reverse-high clutch cylinder are already assembled).

16 Examine Subassembly

Take a close look at this subassembly to verify the order of components from top to bottom–forward planet carrier, forward clutch ring gear/hub, forward clutch, and reverse-high clutch.

Servo Pistons and Seals

Servo pistons and seals call for close attention to detail. Not only is the intermediate servo cover and piston type important to band bite, it is also important to release timing. Transmission professionals I've worked with stress the importance of using C, H, or R servos not only due to drum grip, but also for quicker release time. The preferred servo of choice is the R servo, which offers the quickest wide-open-throttle release time of all the large servos and the obvious advantage of size. Release time is determined by release passage size. Less flare means more snap.

The low-reverse servo piston and seal are one integral bonded assembly, which should be replaced without exception. Both servos and bands should be fastened with new adjustment nuts, which normally arrive in a transmission kit including the B&M Transkit. ■

17 Install Shell & Sun Gear

Input shell and sun gear go on top, mated to the forward planet and reverse planet carrier.

18 Dress Case Mating Surfaces

Dress the transmission case mating surfaces with a stone to achieve a perfect surface to prevent leakage. As surfaces are stoned, irregularities appear.

19 Replace Low-Reverse Servo Piston

The low-reverse servo piston should always be replaced. Check for proper fit, then lube for installation. Install it smoothly until seal is submerged. Be careless here and you not only damage the seal, but also will have to replace the piston.

20 Replace Low-Reverse Servo Cover Seal

Low-reverse servo cover gets a new seal, which should be lubed with transmission assembly lube. Slowly run the bolts down until the cover is fully seated, and then torque to 12 to 20 ft-lbs in crisscross fashion.

21 Replace Intermediate-Band Servo Piston Seal

Intermediate-band servo piston gets new seals and plenty of transmission lube to ensure proper sealing and operation. Lube the seals prior to installation for best results. Best servos are H or R types for full-size Fords and trucks, which are the largest and provide the greatest apply area for better band hook up. They also offer the quickest release time, which is important to crisp performance. The C servo you often hear about is for the 289 High Performance and is harder to find. The C servo is also available as a reproduction from Scott Drake Reproductions if you're determined to have one.

Professional Mechanic Tip PRO TIP

22 Prepare Intermediate Servo Cover and Piston

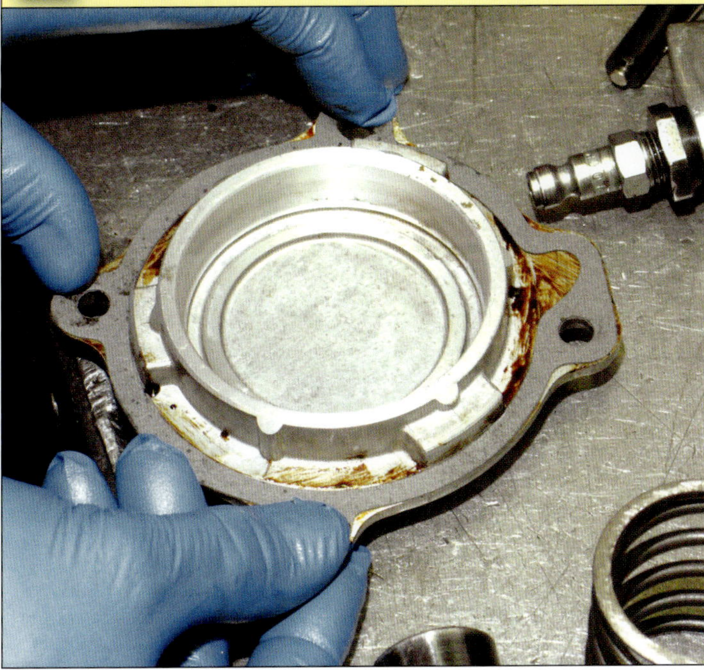

PRO TIP *Steve Bunch of Leon's Transmissions suggests using a super-thin film of non-hardening Permatex Form-A-Gasket for optimum sealing of the intermediate servo cover, though most transmission techs never use sealer. The problem with C4s is leakage, which calls for extreme measures. In some instances, it is acceptable to use sealer, but only a non-hardening type and sparingly. If it comes out around the edges, you're using too much.*

PRO TIP *Gently press the intermediate-band servo piston into the servo cover. Be very careful about seal lubrication and protection. It is easy to unknowingly damage seals and then be faced with leakage when the unit is in the car. Save time by slowing down and getting this right.*

23 Apply Assembly Lube

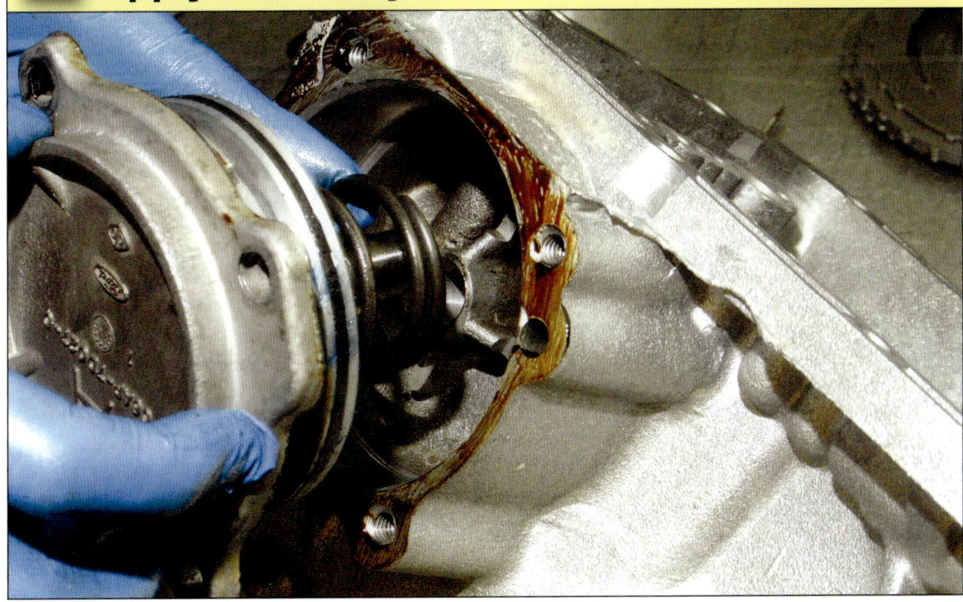

Install servo cover, piston, and spring with transmission assembly lube. Torque cover bolt to 12 to 20 ft-lbs in a gentle crisscross motion until tight. Do not overtighten.

Internal Reassembly

1 Begin Main Case Assembly

With all subassemblies complete, you're ready to assemble the main case, which begins with the number-9 thrust washer at the roller clutch. You can see that the roller clutch outer race has already been installed. Lubricate the thrust washer with transmission assembly lube.

2 Install Roller Clutch Spring Retainer

Roller clutch spring retainer (also called a cage) is next. Some incorrectly call this a sprag clutch.

3 Inspect Inner Race

The roller clutch's inner race ultimately splines inside the low-reverse drum once the roller clutch is assembled. It is shown here with the low-reverse drum and hub. The roller clutch (not pictured) allows low-reverse drum rotation one way, but not the other.

Documentation Required

4 Install Inner Race

The inner race fits in the center of the roller clutch. Note how the spring retainer, roller, and spring go together—with only one set installed here. There are 12 rollers and 12 springs. Install the springs as shown here. It is easy to get the springs backward, which results in prompt transmission failure. Spring lip must point toward the outside of the case at the roller and toward the inside opposite the roller. Rollers lock against the rise in the case, which is how this works as a one-way clutch. Like a ratchet in your toolbox, the one-way clutch allows rotation only one way.

TECH TIP

Roller Clutches

Roller clutches bite amateurs and professionals alike because it's so easy to make a mistake here. Don't be in a hurry. Roller clutch malfunction comes from improper assembly—mostly springs installed backward. Pay very close attention to detail here. Even one roller spring installed backward causes transmission failure. ■

5 Install Parking Pawl

The installed parking pawl, return spring, and gear.

6 Install Governor Distributor

Governor distributor is installed next, gently tapping it into place. There are no gaskets or seals here, just a snug interference fit.

Critical Inspection

7 Inspect for Wear

Replace all iron sealing rings. Inspect the governor distributor contact surfaces inside for wear prior to tailshaft installation.

8 Install Tailshaft

Use transmission assembly lube to lubricate tailshaft seals. Install the shaft, gently working them into the governor distributor. A key to reliable assembly is being gentle. Force stubborn pieces together and you damage critical surfaces and parts.

9 Install Low-Reverse Band

First install the low-reverse band, the large band, which has been soaked in transmission fluid. Follow transmission case provisions for clearance and the low-reverse band goes right in. Expect some stubborn interference issues here. Be patient and it drops right into place.

Low-reverse band and drum look like this when properly installed. On the right is the parking-pawl mechanism rod and detent spring.

Clutch Steels

Should you resurface drums and clutch steels? It's all a matter of whom you ask. When you buy new clutch steels, they come right out of the package with smooth surfaces. The same can be said for clutch drums. Clutch drums could use some help surface-wise for good bite. Gently sand drum surfaces with fine-grain sandpaper (400- or 600-grit) for good band engagement. ■

10 Install Number-6 Thrust Washer

Reverse planet carrier receives the number-6 thrust washer, which is lubricated with transmission assembly lube. Tabs seat into provisions in the reverse planet carrier.

11 Install Number-7 Thrust Washer

Number-7 thrust washer goes on this side of the reverse planet carrier. Have you inspected reverse planet carrier pinions for excessive wear and oscillation? Now is the time to be absolutely sure about wear.

12 Install Reverse Ring Gear & Hub

Number-8 thrust washer is already installed at the low-reverse drum inside the case. Tailshaft retention C-clip is already installed. Now install the reverse ring gear and hub.

13 Install Reverse Planet Carrier

Fit the reverse planet carrier into the reverse ring gear already installed in the case. Reverse planet drives the reverse ring gear and tailshaft. Tailshaft is splined into the reverse ring gear hub.

14 Install Input Shell/Forward Clutch

Forward clutch, reverse-high clutch, and input shell are already assembled and ready to be fitted into the reverse planet carrier. The input shell's sun gear fits into the reverse planet carrier. You have to rock this assembly back and forth to achieve proper planet carrier mating.

15 Install Intermediate Band

Fit the intermediate band around the reverse-high clutch drum. Although this intermediate band is dry, soak it prior to installing it. With street/strip applications, this is the friction material you want (in red); it offers longevity. Blue and Kevlar racing frictions offer outstanding hookup, but a short lifespan.

16 Install Front Pump

With a long punch carefully guide the front pump into place; to where the stator support slides into the forward clutch. Did you remember to install the gasket? Some C4 and C5 applications have an O-ring instead of a gasket. Make sure the O-ring is thoroughly lubricated.

TECH TIP

Governor Screen

Don't forget the governor filter even if your C4 core didn't have one during teardown. The governor screen is tiny and doesn't seem that important. However, it is there to keep debris, such as metal particles or friction material, out of the governor assembly. Just because everyone else leaves this filter out doesn't mean you should. ■

Torque Fasteners

17 Torque Bolts

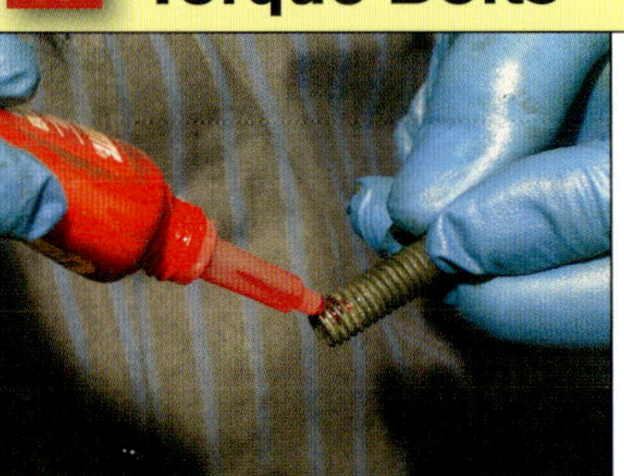

Use a high-temperature thread locker on pump bolt threads for added security. Torque bolts to 28 to 40 ft-lbs in crisscross pattern for even torque.

External Assembly

1 Install Governor Screen

Install the governor filter (screen). It keeps stray particles from adversly affecting governor operation. Do not forget this piece, even if your C4 didn't have one to begin with. You can get one from Transtar or just about any local transmission-parts supply house.

2 Install Governor

The governor installs on the tailshaft. Note primary (right) and secondary (left) valves. Primary valve goes to work at 10 mph to get things started. Secondary valve determines shift points based on vehicle speed.

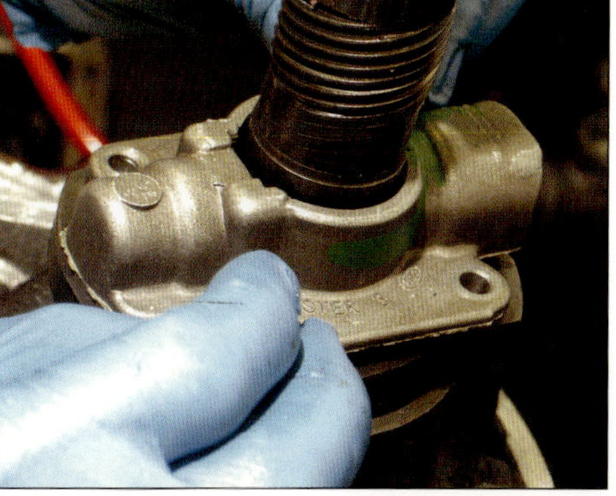

Install the governor and torque bolts to 80 to 120 in-lbs. Note that is inch-pounds, not foot-pounds. Applying too much torque distorts the governor, rendering it useless.

3 Apply Yoke Seal

Apply a thin film of Permatex Form-A-Gasket non-hardening yoke seal to the tailshaft. This is old technology, but still a very effective sealer.

TECH TIP: Throttle Valve

There are two types of throttle valves: screw-in and press-in. Early C4 units prior to 1972 (left) are screw-in with a gasket. Later units are press-in (right) with a hold-down clamp and an O-ring. ■

TECH TIP: Don't Forget Throttle Valve Control Pad

Did you forget something? Throttle valves work best when the control rod, which connects the throttle valve with the valve body, is installed. This is an easy part to forget and improperly install. Check to make sure there is contact with the valve body. C4 transmissions get off to a bad start when this critical part is forgotten–and it is easy to forget. ■

Begin throttle valve installation with the control rod, installed with a magnet. Control rod connects the throttle valve (vacuum modulator) with the valve body.

4 Install Intermediate Band Adjustment Screw

Install the intermediate band adjustment screw along with the locknut. C4 band adjustments are all coarse-thread. C5 intermediate band adjustment is fine-thread, while low-reverse is coarse-thread.

5 Install Struts & Intermediate Band

Here you get a look at the low-reverse band, struts, servo rod, and adjustment screw. Install the intermediate band between the low-reverse band and the servo rod.

Professional Mechanic Tip

6 Adjust the Bands

Tighten in this sequence: intermediate band to 10 ft-lbs torque and then back off 1½ turns. Low-reverse band to10 ft-lbs and then back off 3 full turns. Locknuts to 35 to 45 ft-lbs. Not every transmission shop adjusts bands this way. Many have developed their own technique through the years, based on what works well for them, but this is what Ford suggests. Make sure adjustment screws do not turn when you're tightening locknuts. Locknuts have integral seal lips to keep fluid inside.

TECH TIP

Compressed Air

Air-check all components before buttoning up your C4. Compressed air works the same way as transmission fluid, only without the mess. Determine proper passages and give it a whirl. With each passage, you are able to hear the affected component. You are able to hear clutches and servos operate as air is applied. ■

Air-check the intermediate band servo. When air is applied, the intermediate band servo applies the band. Air-check all transmission hydraulic components before closing up.

Final Assembly

1 Replace Shaft Seal

PRO TIP Manual-shift shaft passage gets a new seal. Make sure this seal gets generous lubrication with either transmission fluid or assembly lube before shaft installation. Shaft nut is tightened with a 7/8-inch open-end wrench.

2 Install Valve Body

Manual-shift valve (bottom) is tied to manual-shift control linkage. Kickdown valve (top) is depressed by the kickdown linkage at wide-open throttle.

When installing the valve body, carefully guide manual shift and kickdown linkages into proper positions. Visually ascertain proper location and function of each before closing up.

3 Install Transmission Pan

Apply a very thin film of Permatex Form-A-Gasket (non-hardening) between the case and the gasket and allow to set before you install the pan. This is one key to a leak-free installation. You want a pan that's distortion free, which means no irregularities. (New stamped-steel pans are available from Scott Drake Reproductions dealers.) If you want fluid to run cooler, go with a cast finned aluminum pan. The downside: cast aluminum pans tend to bleed fluid through casting irregularities. Although I've seen GE Glyptol used on the inside of cast aluminum pans, some brands of transmission fluid can break down the Glyptol.

Important!

4 Install Converter

Install the converter (here a B&M Holeshot with 2,400-rpm stall speed) and take extra care to properly seat input shaft and pump flats. Proper fitment boils down to feel—first the input shaft, and then the pump flats. If you can fit your fingers between converter and front pump, the converter isn't seated properly. Before installing the converter, pour at least 1 quart of transmission fluid into the converter to help prime the system for generous lubrication and pressure on start-up.

One mistake I see made time and time again is not replacing all bushings. As a matter of practice and economics, most transmission shops don't replace the bushings unless they're damaged, but this is a bad practice. Because bushings do most of your C4's work, you should replace them even if they don't look worn. Performance Automatic has complete bushing kits for C4 transmissions.

Although six-pinion forward planets are for racers, they're also good for street C4s because they provide long-term durability—more gear teeth to share the load, making your street C4 super-rugged. This is especially important if you're going to tow or go weekend racing.

Performance Automatic has this larger intermediate band servo for C4 transmissions. Because there's more piston and bore surface area, count on a more solid band application and quicker release time.

Hardened input shafts from Performance Automatic are available for 1971-on C4 transmissions. These are good for racing and towing applications.

C4 Clearances

Function	***Measurement (inch)***
Geartrain Endplay	.008–.042 using selective-thrust washers
Turbine/Stator Endplay	.023 maximum new
	.040 maximum used
	.044 maximum new
	.060 maximum used
Intermediate Band Adjustment	Torque adjustment to 10 in-lbs, then back off 1½ turns; tighten locknut to 35-45 ft-lbs
Low-Reverse Band Adjustment	Torque adjustment to 10 in-lbs, then back off 3 turns; tighten locknut to 35–45 ft-lbs
Selective Snap-Ring Thicknesses	.102–.106, .088–.092, .074–.078, .064–0068
Forward Clutch Clearances	.020–.036 or .026–.042 depending on year and type
Reverse Clutch Clearances	.050–.056 for all

Cast aluminum pans not only look sharp, but also provide additional cooling capacity. Although Performance Automatic provides a drain plug, always remove the pan and change the filter while you're on your back doing transmission service. Changing fluid alone is never a good idea.

Performance Automatic tends to be at the center of the C4 universe because the company has done so much racing research and development in nearly four decades of racing experience. Performance Automatic can sell you a complete C4 ready for street or racing, and it can freshen your C4 core as needs warrant.

Torque Converter Stall Speeds

Engine	*RPM*
170-ci Six	1,450–1,650
200-ci Six	1,600–1,800
240-ci Six	1,330–1,500
289-2V V-8	1,750–1,950
289-4V V-8	1,800–2,000
302-ci V-8	1,450–1,600
351W-2V V-8	1,580–1,780
351W-4V V-8	1,650–1,850

Torque Specifications

Location	*Foot-Pounds*
Torque Converter to Flexplate	23–28
Bellhousing to Case	28–40
Pump to Case	28–40
Roller Clutch Race To Case	13–20
Oil Pan	12–16
Stator Support to Pump	12–20
Intermediate Servo Cover	16–22
Low-Reverse Servo Cover	16–22
Diaphragm to Case	15–23
Governor Distributor to Case	12–20
Manual Lever to Shaft Nut	35–45
Tailshaft Housing	28–40

Location	*Inch-Pounds*
Valve Body End Plate	20–35
Valve Body Half Bolts	40–55
Filter-to-Valve Body Screws	40–55
Filter-to-Valve Body Bolts	80–120
Filter/Valve Body-to-Case Bolts	80–120
Neutral Safety/Backup Light Switch	55–75
Governor-to-Output Shaft Bolts	80–120
Cooler Line Fittings	80–120

C4 Shift Kits

To improve shift quality in Ford's venerable C4, you can look to B&M Racing & Performance, which has been producing some of the best transmission performance parts in the world for more than 50 years. The B&M story dates back to 1953, when Bob Spar and Mort Schuman put their heads together and founded not only a great company, but also an institution for the automotive masses that's still going strong. Their first effort was known as the B&M Hydro Stick 4-speed automatic.

B&M's vast experience with automatic transmissions includes the Ford C4 Transpak kit (PN 50227), which is for 1964–1966 C4 Dual-Range Green Dot transmissions, a rare treat for these early first-generation C4s with their unique valve bodies. The kit comes in three basic configurations: Heavy Duty, for towing; Street, for cruising and drag racing; and Competition, for all-out drag racing. While almost everyone else in the transmission business has given up on the Dual-Range C4, B&M has stayed with it, making firm shifts and longevity possible for these first generation C4 units.

Although I refer to this B&M Transpak kit, my main focus is the more-common 1970–1982 C4 transmission and the Shift Improver Kit (PN 50262) available for these slush boxes. This B&M kit is engineered to work with all 1970–1982 C4 transmissions. To make the most of a Shift Improver Kit, you have to start with a healthy transmission—with fresh clutches, bands, and seals.

B&M suggests allowing your C4's transmission pan to cool down before removal because these units operate at temperatures close to 250 degrees F. A cool-down period also allows contaminants to settle in the sump before removal and drainage. Although it's a good idea to change all transmission fluid, it's not always good to drain the torque converter because you run some risk of pump cavitation and a dry startup, which you do not want.

Another thing to prevent is fresh fluid "shocking" the seals. This means you want some of the old fluid (as long as it isn't burned or heavily contaminated with clutch/band material) mixed in with the new. Though this has been suggested to me by transmission professionals, it may not be gospel. But you do want to keep a certain amount of fluid in the converter to keep the pump primed if for no other reason.

B&M Kits for C4s

The B&M Transkit is more than just a shift improvement kit. In fact, it does everything the B&M Shift Improver Kit does and allows you to downshift or upshift manually at any speed. This gives you great flexibility as long as you handle that flexibility responsibly.

Transkit PN	*Years*
50227	1964–1966 C4 Dual-Range
50228	1967–1969 C4
50229	1970–1982 C4

For those who want little more than a firmer shift, the B&M Shift Improver Kit is perfect for your C4. The Shift Improver Kit is not available for the first-generation Dual-Range C4. However, it is available for the 1967–1982 C4, which encompasses most of the C4 transmissions out there. The Shift Improver Kit not only firms your shifts, it improves transmission life by reducing clutch and band slippage.

Shift Improver Kit PN	*Years*
50260	1967–1969 C4
50262	1970–1982 C4

Disassembly

1 Remove Bolts & Drain Fluid

After your C4 has cooled down, carefully remove pan bolts and allow fluid to drain. Observe fluid color and consistency. If it is burned (brown in color), you need more than a shift improvement kit. Fluid should be pink and smell fresh. Burned fluid is a sign of deeper troubles and the need for a teardown.

Critical Inspection

2 Examine Filter

Remove filter and examine screen for contaminates (friction material). Take note of bolt locations and lengths. Fluid here looks pink and clean indicating a healthy C4 transmission. Valve body removal involves several bolts using a 3/8-inch socket.

3 Remove/Replace Throttle Valve

With each transmission service, replace the throttle valve, also known as the vacuum modulator. It's a good idea to remove the throttle valve and rod prior to valve body removal. Keep close track of the actuating rod because it's easy to misplace.

4 Prepare Shift Kit

This is B&M's Shift Improver Kit (PN 50262) for 1970–1982 C4 transmissions. You may go with this simple kit or the more complete Transpak (not shown), which includes filter, throttle valve, springs, and valve parts. Either way, follow B&M's instructions carefully.

5 Use New Filter & Gasket

When you install any B&M kit, you should also install a new filter and gasket for best results.

Documentation Required

6 Remove Valve Body Bolts

There are two halves to the C4 valve body. Remove all bolts using a 5/16-inch socket, taking note of where each bolt goes. Remove the manual shift detent last.

Important!

7 Remove Drain-Back Valve

The torque converter drain-back valve comes free when the filter is removed. One mistake I hear about more than any other is this check valve being lost during C4 service. Whenever you're replacing the filter, don't forget this very important part, which is easily lost in a drain pan.

8 Remove More Bolts

Use a 5/16-inch socket to remove the rest of the bolts on the top half.

9 Rotate Valve Body

Before continuing, flip valve body over. This prevents check balls from being lost.

10 Remove Separator Plate

Remove the separator plate from the bottom half.

Documentation Required

11 Note Location of Check Balls

On the left is the bottom half of the valve body with filter. On the right is the upper half with valves. Both halves have check balls, which should be properly located and noted during disassembly. B&M's instructions are very specific about which balls go back and which ones do not.

12 Remove End Plates

Remove the end plate for the transition valve, 2-3 back-out valve, and cut-back valve.

13 Remove Valves

Carefully manipulate and remove back-out, transition, and cut-back valves.

Professional Mechanic Tip

14 Replace Springs

When you remove the back-out, cut-back, and transition valves you may find that the springs have already been discarded during a previous performance buildup and two 7/32-inch check balls were installed. B&M suggests discarding the back-out and transition-valve springs as part of the PN-50262 modifications. One 7/32-inch ball and the brass disc in the kit do the same job as these balls.

Reassembly

1 Reinstall Valves

Reinstall the back-out, cut-back, and transition valves. Valves must be flush.

Professional Mechanic Tip

2 Reinstall End Plate

Reinstall and torque the end plate; Ford suggests 20 to 30 in-lbs. B&M prefers10 to 15 in-lbs, at the most.

3 Install Check Ball (If Applicable)

A 7/32-inch check ball is installed here on all except Pinto, per B&M's PN-50262 instructions. Discard the original rubber disc (arrow).

Critical Inspection

4 Examine Separator Plates

Examine both separator plates to determine a proper course of action. B&M suggests laying plates on top of each other to determine compatibility. In other words, make sure the B&M plate matches your C4's valve body.

5 Drill Separator Plate

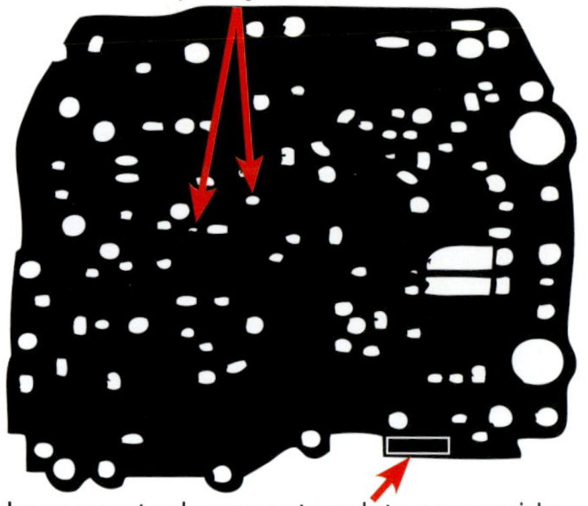

Use your C4's separator plate as a guide for drilling and follow this illustration, which applies to all C4 applications except Pinto in 1970–1982.

6 Drill Pinto Plate (If Applicable)

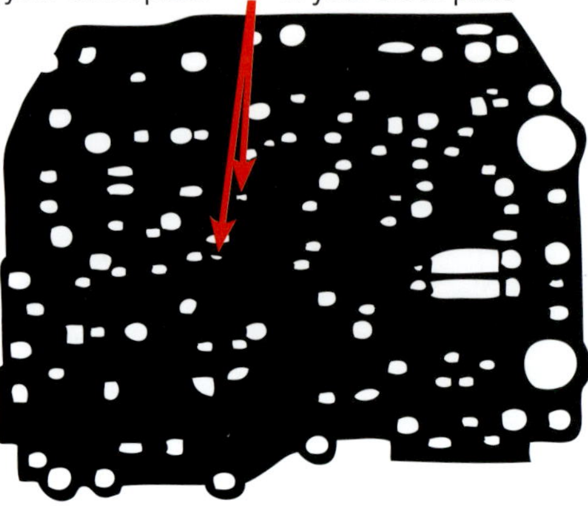

If you have a Pinto C4 separator plate, drill per this illustration.

7 Drill Pinto Plate (If Applicable)

If you have a street/strip application, drill two 3/16-inch holes per B&M's instructions. After these holes are drilled, take a larger bit and deburr the edges.

8 Install Check Ball

Install a 7/32-inch check ball here.

Professional Tip

PRO TIP

9 Install Separator Plate

Install separator plate as shown. B&M suggests not using a gasket here.

10 Reunite Halves

Reunite the halves, making sure all check balls are in place. Use transmission assembly lube to secure check balls before assembly.

Torque Fasteners

11 Torque Bolts

In a criss-cross fashion, torque valve-body half bolts to 20 to 30 in-lbs each, per Ford. This might be too much torque, with the risk of pulling threads out.

12 Reinstall Drain-Back Valve

Reinstall the torque converter drain-back valve with valve cup toward the inside.

13 Reinstall Manual Valve Detent

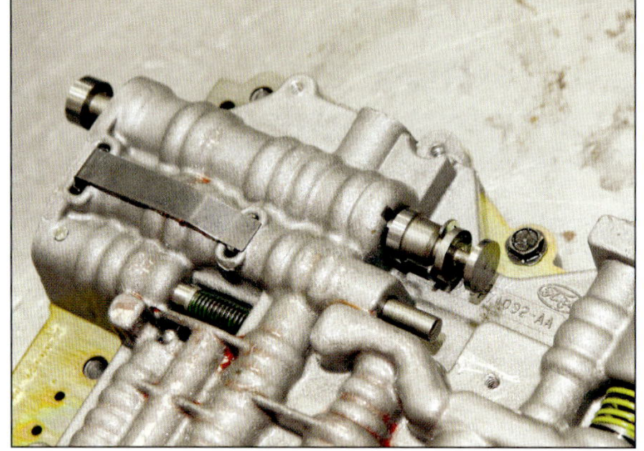

Reinstall the manual valve detent.

14 Inspect Assembled Valve Body

Assembled 1970–1982 valve body should look like this. Remember, there are two basic types of manual shift valves: one with an external detent (shown) and one with an internal detent (not shown). Internal detent is 1964 through the early 1970s. External detent came along in the 1970s. Make sure you don't get these two types mixed up.

15 Note Shift Valve and Kickdown

A closeup of the manual shift valve (top) and the kickdown (bottom).

16 Note Detent

An earlier C4 valve body with internal manual valve detent, which is quickly identifiable by the manual valve detents in the valve itself. External detent is smooth.

17 Inspect Transpak Kit

This is B&M's Transpak kit (PN 50227) for 1964–1966 C4 Dual-Range transmissions.

18 Drill Separator Plate

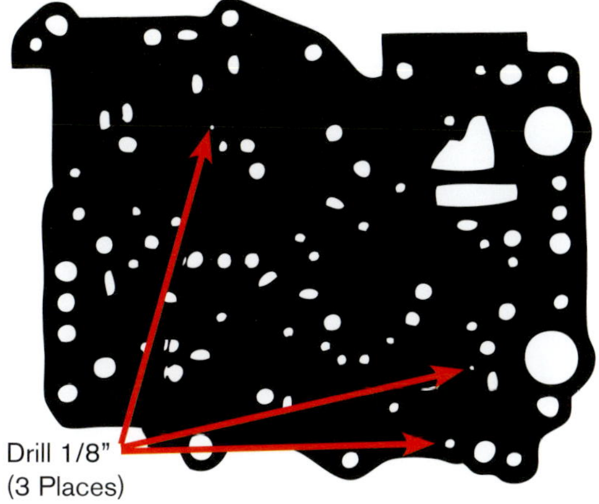

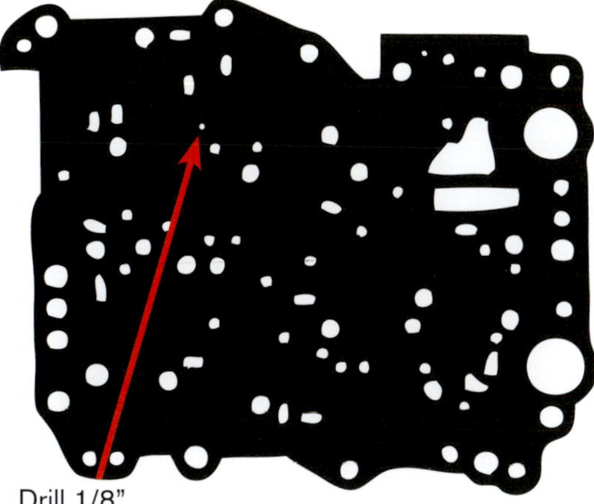

If you have the PN-50227 kit, use the pattern on the left to drill the separator plate for 1964 Dual-Range only. For 1965–1966 C4 Dual-Range, follow the pattern on the right.

Professional Mechanic Tip

PRO TIP

19 Avoid Leakage

PRO TIP *Because C4 pans are notorious for leakage, go with a new reproduction pan or dolly yours straight. Leon's Transmissions suggests a thin film of Permatex Form-A-Gasket (nonhardening) on the pan side of your gasket. Not all transmission professionals agree with this approach, but it works.*

20 Adjust Throttle Valve

After you've properly reinstalled the valve body with manual and kickdown valves correctly positioned, install the new B&M throttle valve. Adjustment calls for turning clockwise to increase pressure and counterclockwise to decrease pressure. One full turn clockwise increases control pressure by 2 to 3 pounds, and counterclockwise reduces pressure by 2 to 3 pounds.

CHAPTER 6

C6 Select-Shift

Ford's C6 transmission was first pressed into service for 1966. It was designed to replace a generation of heavy and outdated BorgWarner MX and FX iron-case transmissions used behind Ford big-block and Y-Block V-8s. The C6 is on a par with GM's Turbo-Hydramatic 400 and Chrysler's 727 Torqueflite transmissions. It is a heavy-duty 3-speed automatic transmission designed primarily for high-power V-8 engines and employs a conventional, non-locking torque converter.

Internal C6 architecture isn't much different than that of the smaller C4 introduced in 1964 (see Chapters 4 and 5). The C6 differs in its size and having a larger case—where bellhousing and main case are one integral casting. This means you are faced with choosing between at least four basic types of C6 bellhousing patterns. Here's what you can expect to find and how to differentiate between them:

Engine	*Bellhousing*
FE Series Big-Blocks	Round six-bolt
385-Series 429/460 and 351/400M	Angular six-bolt (with closed bottom)
6.9L/7.3L Diesel Engines	Six-bolt similar to 385-Series (with open bottom)
Small-Block	Six-bolt

The C6 consists of a torque converter, front hydraulic pump, Simpson compound planetary gearsets controlled by one band, three disc clutches, a single one-way roller clutch, and a simple mechanically controlled hydraulic control system. The C6 differs from the smaller C4 in its low-reverse function. Instead of a low-reverse drum as in the C4, you instead have a low-reverse clutch package tied to the case, which performs the same function in a C6. This means the C6 is equipped with one band and servo instead of two.

For its first year, 1966, the C6 was a Dual-Range Cruise-O-Matic Green Dot transmission like the C4, where normal driving takes place on the green dot at the detent. For special driving conditions on snow and ice, the small dot (off detent) starts you out in second gear, with upshift into final drive once you get going. For the first-year the C6 was a stand-alone unit, meaning the valve body was Dual-Range for that year only.

For 1967, the C6 became Select-Shift with a routine P-R-N-D-2-1 shift pattern just like the C4. For slippery conditions, you start out in "2" or second gear to get going. To get an upshift into final drive, you must move the shifter into "D" or Drive.

Aside from minor changes in the 1970s and 1980s, the C6 didn't change much over its long production life from 1966 to 1996. The C6 remained in production long after it was dropped from Ford passenger cars and trucks because it remained popular with companies that needed it for heavy-equipment applications.

This is the small-block C6 with the six-bolt small-block bellhousing for 351W and 351C engines. You may also use this transmission behind your 289/302 V-8, though I don't suggest it. There's simply too much weight and parasitic drag going on here to be used behind the lower-displacement V-8s. Venting comes from a mushroom vent on top of the case.

The C6 for 385-series 429/460-ci big-blocks and 351/400M engines. Don't get this one confused with the C6 for Diesel engines, which is the same but with a cutout at the bottom of the bellhousing for torque converter access. Some 385-series and M-series C6s are finned and others are not. Fins add strength and some degree of cooling capacity because they radiate heat.

C6 Operation

C6 function isn't much different than you find in the C4 with the exception of having a low-reverse clutch instead of a low-reverse band and clutch drum. Function begins with the front pump, a simple gear pump that provides hydraulic pressure for operation and lubrication for moving parts. The system is protected with drain-back and pressure-relief valves. The drain-back valve keeps the torque converter full after shutdown. A pressure-relief valve prevents overpressure and catastrophic failure.

The C6's power flow follows the same path as the C4's except for the low-reverse function. You get better load distribution and torque multiplication through Simpson planetaries and a ring gear around a sun gear. And this is where you are in first and second gear.

The objective is to get power from the input shaft to the output shaft via a series of planetaries and clutches. The C6's input shaft is splined into the forward clutch. In neutral, the forward clutch turns with the input shaft and torque converter turbine. With all of its clutches

Core Choices

When you're searching for a C6 core, be mindful of vintage and where the core originated. Which valve body does it have? The C6 Dual-Range valve body (1966 only) is the Green Dot design and not desirable for performance use. In fact, it is hard to find a shift improvement kit for the Dual-Range. Opt for 1967-on. Closely examine Ford casting numbers on both valve body halves to determine origin. ■

This rounded bell is the FE-series big-block C6, which doesn't have fins as on the 385-series/M-Series and small-block C6s. Some FE C6s were finned later in production for added strength.

Ford C6 castings are identified by Ford casting numbers like this one—D5UP-7006-AA—on a mid-1970s main case.

Expect to see Ford casting and part numbers on all castings including: tailshaft housings, clutch pistons, ring gear hubs, governors, and more. This makes these items easier to identify, but it isn't a perfect system. The year noted isn't always when the part was cast. This number is only the last engineering change, so expect to see a 1977 that was actually cast in 1979 as one example. And because the valve body is in two halves, expect to see different casting number years on each half, depending upon engineering changes.

and steels, the forward clutch transfers power to the ring gear, which is mated to the forward planet carrier. The forward planet carrier is tied to the output shaft via the sun gear. When the transmission is in first or second gear, the forward planet is always in motion.

Through the Gears

As the reverse-high clutch (direct clutch) and input shell come into play, power begins to flow. The reverse-high clutch rides on the pump stator. It is also mated to the forward clutch. When you apply the reverse-high clutch, it engages the forward clutch. The reverse-high clutch is linked to the input shell, which is connected to the sun gear. As power flows to the sun gear, it travels to the low-reverse or rear planet carrier to the low-reverse ring gear, which is splined to the output shaft.

For the most part, I've described C4 function to this point with the same cast of players. The C6 differs in where the power travels at this point. Instead of a low-reverse band and clutch drum, you have a low-reverse clutch locked into the transmission case. When you apply the intermediate servo and band to the forward clutch cylinder, power is channeled through the forward planet, which becomes second gear during the 1-2 upshift.

In first gear, the input shaft turns the forward planet, which reverses input shaft rotation to the sun gear. The sun gear is splined into the low-reverse (rear) planet, which again reverses rotation to the output shaft. This gives you a gear reduction of 2.46:1 (2.46 revolutions to every 1 revolution of the output shaft).

In second gear, fluid under pressure energizes the intermediate servo, thereby locking the band, forward clutch drum, and sun gear. The forward planet rotates around the sun gear, which provides support at this point. This drives the forward ring gear, which drives the output shaft in the same direction at a ratio of 1.46:1.

When a C6 shifts into final drive, the intermediate band releases the forward clutch drum, and both forward and reverse-high (direct) clutches are applied. With the reverse-high and forward clutches applied, both input and output shafts are connected and the entire package rotates at the same speed for a 1.00:1 ratio.

In reverse gear, the low-reverse clutch is applied to hold the reverse planet stationary. When this happens, the reverse-high clutch is applied to get the input shaft tied to the sun gear. Because the reverse planet is held still, this causes the sun gear to be driven in the opposite direction as the input shaft. This turns the the ring gear, which is tied to the output shaft, at a ratio of 2.175:1.

At a glance, the huge C6 pan has a little dogleg at the vacuum modulator, which makes it easy to identify.

Pressure Relief Valves

Hydraulic pressure is there for control and lubrication purposes in the C6. Pump pressure, also known as line or control pressure, has to be controlled for proper and safe operation. Fluid flow begins at the pump cavity and flows into the torque converter. This is why it is important to have fluid in the converter during installation. Fluid in the converter primes the system and gets fluid moving from the sump to the pump to the converter. It's a momentum thing.

There are two valves important to this discussion: the drain-back valve and the converter pressure-relief valve. The drain-back valve's job is to keep fluid in the torque converter after shutdown. It unseats at 5 psi during engine start and allows fluid flow. The converter pressure-relief valve, which vents to the sump during overpressure (90 psi), is next in line. The main regulator valve is next, which regulates control pressure.

Next in line is the governor, which is mounted on the tailshaft. The governor has two valves that work hand in hand. The primary valve starts to work when vehicle speed reaches 10 mph. This keeps the governor from going to work below this speed. Once the primary governor opens, allowing fluid pressure to pass, the secondary valve comes on line as a means of controlling the shift point based on vehicle speed.

Valve piston movement is countered by spring pressure. As vehicle speed increases, centrifugal force causes the valve pistons to override spring pressure and the valves open one at a time with corresponding speed.

Throttle Valve Operation

Shift control is determined by both the vacuum modulator (throttle valve) and the kickdown linkage after the governor comes on line. Kickdown linkage is for wide-open-throttle operation. As with the C4, the C6's vacuum modulator ties manifold vacuum to transmission control pressure. The greater the intake manifold vacuum, the lower the control pressure for a sooner, softer upshift. When we have low intake-manifold vacuum at wide-open throttle, the throttle valve helps increase control pressure for a firm upshift, which is what you want at wide-open throttle.

As with the C4, throttle valve adjustment on the C6 is key to proper operation and reliability. You've got to get this right or face slippage and transmission failure. To adjust the throttle valve properly, you've got to know control pressure, which involves a pressure gauge at the control pressure port just above the manual shift and neutral safety switch. Transmission sump temperature should be at a hot idle. Cold transmission fluid gives you an erroneous pressure reading on the high side because the fluid is denser when cold. The gauge pressure range needs to be 0 to 400 psi.

First, you must establish proper throttle valve function. Does the throttle valve work properly? Is there vacuum leakage? This is checked with the throttle valve removed using a vacuum pump or intake manifold vacuum from an operating engine at idle. At 18 to 22 inches of vacuum, you are able to tell if there's leakage by a hissing sound or the absence of diaphragm/rod movement. There should also be diaphragm and rod movement at 18 to 22 inches of vacuum. At higher elevations, your engine may struggle to maintain 18 inches at idle.

With the throttle valve installed, transmission in neutral/park, and engine at idle, you should have a minimum of 18 inches of manifold vacuum. Anything lower than 18 inches of vacuum indicates engine health problems that must be corrected or you have a really lumpy

C6 Function

Gear	*Ratios*	*Intermediate Band*	*Direct Clutch*	*Forward Clutch*	*Reverse Clutch*	*One-Way Clutch*
Low	2.46:1	Off	Off	On	On	Holding
Second	1.46:1	On	Off	On	Off	Over-Running
Drive	1.00:1	Off	On	On	Off	Over-Running
Reverse	2.175:1	Off	On	Off	On	N/A

C6 Line Pressure Guide

Engine Vacuum (inches)	*Throttle Status*	*Shifter Selection*	*Control Pressure (psi)*
18 and Higher	Closed/Idle	Forward Ranges and Neutral Only	40–61
18 and Higher	Closed/Idle	Reverse Only	60–93
In Transition at 17 and Lower	Opening	Forward Ranges Only	Should show a rise in line pressure
10	Open	Forward Ranges Only	110–114
Low Vacuum	Open	Forward Ranges Only	154–180
Low Vacuum	Open	Reverse Only	234–275

camshaft profile. If you have at least 18 inches of vacuum and the selector is in any forward gear, you should see 40 to 61 pounds of line pressure. In reverse, you want 60 to 93 pounds of line pressure.

With the throttle open at 10 inches of vacuum and the transmission in any forward gear, you should see 100 to 114 pounds of line pressure. At low vacuum, expect to see 154 to 180 pounds of line pressure in any forward gear. In reverse at low vacuum, expect to see 234 to 275 pounds of line pressure.

Whenever line pressure isn't within these parameters, throttle valve adjustment must be done with great care. When line pressure is too high, shifts become harsh. If line pressure is too low, shifts become soft (slippage) and that's when you do damage with burned clutches and bands. To increase control pressure, throttle valve adjustment needs to be turned clockwise. To reduce line pressure—adjustment is counterclockwise.

Ford says one full turn either way adjusts line pressure 2 to 3 psi. Check line pressure at idle, at 10 inches, and then at 3 inches of manifold vacuum to determine if any further adjustment needs to be made. Never base your adjustments on shift-feel alone; that isn't what line pressure is all about. Always use a pressure gauge and follow Ford's guidelines to the letter.

C6 Line Pressure Quick Reference

Intake Manifold Vacuum at Idle (inches)	*Line Pressure (psi)*
17	55–68
16	55–75
15	55–81
14	55–87
13	55–94
12	55–100
11	55–108

Intermediate Servo and Apply Levers

Servo selection for the C6 isn't nearly as involved as it is for the C4 because there really is only one type, the R servo, to ensure good intermediate band hook-up and release. But you must be careful to properly match servo covers, pistons, and apply levers. This requires thorough examination of the Ford Master Parts Catalog if you want to be absolutely certain of the intermediate servo pieces you need for your C6.

Expect to find N, L, P, and S servo covers, which are the most common servo covers, though I am sure there are more. The transmission professionals I've consulted with all agree on the R servo or an equivalent aftermarket billet replacement. Where it can get tricky is the variety of intermediate servo apply levers.

If you're going for street/strip performance, you should investigate the 6061-T6 billet-aluminum intermediate servo apply levers available from Sonnax Performance, which offer durability as well as mechanical advantage. Three ratios are available: 1.85:1 (E-ratio), 2.30:1 (F-ratio), and 2.82:1 (FF-ratio).

That way, you know exactly what you have when it's time to build. And the R servo works well with any of the stock apply levers I've seen in a variety of builds. In other words, the R servo is a no-brainer because racers have been using it for decades with great success. The R servo arrived for the C6 with the 428 Cobra Jet engine in 1968 and is a proven performer.

The C6, unlike the C4, uses an intermediate servo apply lever to help provide mechanical advantage to apply the intermediate band. There are six servo-apply-lever ratios to be aware of, and you must be very careful of which ones you use with corresponding servo covers and pistons. Broader Performance advises against the use of the F 2.30:1-ratio apply lever and R servo because shifts become too aggressive and can damage your C6.

Here are the six known Ford servo-apply-lever ratios. The higher the servo-apply-lever ratio, the greater the apply force. These letter codes can be found cast into each apply lever.

Lever Code	*Gear Ratio*	*Lever Code*	*Gear Ratio*
A	1.65:1	E	1.85:1
B	1.73:1	H	2.18:1
D	1.97:1	F	2.30:1

TECH TIP

Vacuum Modulators

When ordering transmission parts for your C6, remember there are two basic types of vacuum modulators for the C6: screw-in and press-in. There's also a high-altitude type in both screw-in and press-in versions. Screw-in types are from 1966 to 1971. Press-in types are 1972-on with a retaining bracket. Control rods vary in length depending on what you want your C6 to do. ■

Apply Levers

Here's what you need to know about apply levers from Sonnax Performance:

Lever	*Ratio*	*Part Number*
E, C6AZ-7330-D	1.85:1	31916E
F, C6AZ-7330-E	2.30:1	31916F
FF, Aftermarket	2.82:1	31916FF

Sonnax's FF apply lever does not have a Ford equivalent because it is a very aggressive ratio for racing only. Sonnax also stresses not using the FF with the H or R servo because operation becomes too aggressive. You risk shattering the transmission case if you use the FF servo apply lever with the H or R servo. Use the FF with the appropriate servo. According to Sonnax, the E and F levers work quite well with the R servo.

If you're not wild about one of the billet R servo kits, Transmission Parts USA has a cast aluminum R servo kit (PN K36528R) for your C6 project. This kit is complete with cover, piston, springs, and seals.

Front Pump

Protocol for the C6 front pump is the same as it is for the C4, with no special modifications needing to be made. Inspect the pump cavity and gears for abnormal wear patterns and scoring. Any excessive wear patterns call for pump replacement. TCI Automotive has provided Tom's Transmissions with a remanufactured front pump assembly ready for installation, although I suggest you never install a component right out of the box without close inspection. And then fill the pump cavity with assembly lube for a good start-up.

Valve Body

Aside from that first year with the Green Dot Dual-Range C6 pattern, the C6 valve body didn't change much over its 30-year production life. The first-year Green Dot valve body is discouraged unless you're restoring a stocker and originality is important to you. During the 1970s, the manual valve went from an internal detent to an external detent with a case-mounted detent roller and band. This is something to be mindful of as you amass parts for your C6. Make sure your valve body and transmission case match. If you have the external detent mounted on the case, you want a free-sliding manual valve. If there's no external detent, you want the valve body with an internal detent with a ratcheting manual valve.

Whenever you source a C6 valve body, always determine origin because applications vary greatly. For example, you wouldn't want a C6 valve body for a Diesel application for your 390 or 428 because shift programming and overall operation is quite different due to different pressures and calibrations. Diesels call for an extremely firm shift program for a more solid transfer of torque.

Broader Performance offers a reasonably priced variety of C6 valve body packages depending on how you intend to use your C6. Broader Performance goes to extreme detail with its blueprinted valve bodies, offering you everything from a stock auto-shift valve body, manual and reverse-pattern, to pro tree. These valve bodies are machined to provide a perfect mating surface. Then, they're ultrasonic cleaned before assembly to ensure all debris is removed. Each C6 valve body is assembled and calibrated to your application.

The main thing to remember when searching for a valve body core is to know where it came from for reasons of compatibility and calibration. Read casting numbers and date codes, then, check the Ford Master Parts Catalog to determine origin.

Building Tips

When Ford designed the C6 to begin with, it didn't mess around. This transmission was designed and produced to be Ford-tough like no other automatic had been to date. This means the C6 takes a pounding with stock components and red street frictions and comes back for more. That's the good news, and there isn't much bad. This is an extremely rugged transmission as it came from Ford. For your street build, few modifications are necessary to achieve durability and performance.

The only really bad news with C6 transmissions is internal drag due to dated thrust washer design, which promotes friction. There's also the excessive weight of those tough internals. Figure on losing 50 to 60 hp just trying to turn a stock C6 transmission, which is why the C4 became such a popular drag-racing transmission. But you didn't read this chapter to learn the C6's disadvantages—you want to know how to build one. All racers have their C6 tricks, and I touch on a few of them here. The objective is to build a sturdy street/strip/tow transmission that serves you well for thousands of miles.

The C6 was replaced in Ford vehicles by the E4OD 4-speed automatic overdrive transmission in 1989, which has internals very similar to the C6 and an added overdrive unit. In fact, the E4OD utilizes extreme-duty pieces that work well in your C6 build once you know what you're doing.

C6 Intermediate Servo Covers and Pistons

When Garrett Marks (of Mustangs Etc.) went to work researching the subject of C6 intermediate servo covers and pistons with me, he discovered more intermediate servo cover, piston, and servo apply levers than could ever be covered in this book. I cover the most common types here without getting into specific applications. Keep in mind, not all C6 cases accommodate the R servo and may have to be machined.

If you want additional information, review the Ford Master Parts Catalog. Unfortunately, sizes are not always listed there, which makes this subject even more confusing. You can also look to the aftermarket, specifically Sonnax, for the right combination of servo covers, pistons, and apply levers. Another excellent source for C6 components and information is Broader Performance.

Intermediate Servo Cover Identification

Identification	*Part Number*	*I.D. (inches)*
"D" C6AP-7D027-A	C6AZ-7D027-A	2.69
"E" C6MP-7D027-A	C6AZ-7D027-B	2.77
"G" C6SP-7D027-A	C6AZ-7D027-C	2.72
"H" C6AP-7D027-D	C6AZ-7D027-D	2.72
"J" D1OP-7D027-A	D1OZ-7D027-A	2.35
"L" C8VP-7D027-C	C8VY-7D027-C	2.35
"N" D3AP-7D027-AA	N/A	2.92
"P" N/A	D4LY-7D027-A	N/A
"R" C8OP-7D027-A	C8OZ-7D027-A	3.15
"S" N/A	D7AZ-7D027-A	2.86
N/A	C8SZ-7D027-A	3.00

Intermediate Servo Piston Identification

Identification	*Part Number*	*Seal I.D. (inches)*
RF C6SP-7D022-A	C6AZ-7D021-A	1.79
RF C6MP-7D022-A	C6AZ-7D021-B	1.69
RF C6AP-7D022-A	C6AZ-7D021-C	1.76
RF C6AP-7D022-D	C6AZ-7D021-D	2.03
RF C8OP-7D022-A	C8OZ-7D021-A	2.18
C8VP-7E221-C	C8VY-7D021-B (replaced by D5TZ-7D021-A)	1.79
RF D1OP-7D022-AA	D1OZ-7D021-A	1.79
D5UP-7E221-BA	D5UZ-7D021-A (replaced by F2TZ-7D021-B)	N/A
D5TP-7D221-CA	D5TZ-7D021-A	N/A
D7TP-7E221-AA	D7AZ-7D021-A (replaced by F2TZ-7D021-C)	N/A
D5TP-7E221-DA	D5TZ-7D021-B (replaced by F2TZ-7D021-A)	N/A
RF D3AP-7E221-AA	D3AZ-7D021-B (replaced by D5UZ-7D021-A)	N/A
D4LP-7E221-AA	D4LY-7D021-A (replaced by D5TZ-7D021-B)	N/A
F2TP-7E221-BA	F2TZ-7D021-B	N/A
F2TP-7E221-CA	F2TZ-7D021-C	N/A

Note: Although servo covers and pistons are side by side, this does not indicate a match. Look to the Ford Master Parts Catalog under "Carline Types" to find factory combinations. Space limitations prohibit some of them from being listed here. These are the most common types of servo covers and pistons you can expect to find out there.

Torrington Bearings

Instead of using thrust washers, these components are designed to reduce internal friction by using Torrington bearings, which are actually upgrades found in the Ford E4OD heavy-duty overdrive transmission. The E4OD transmission isn't much more than a C6 with overdrive, a fresh case, and low-friction components inside. The Low-Drag Planetary gear set offers the durability and mechanical advantage of six-pinion performance instead of three each as you find in the C6. Brian Fotrune of Tom's Transmission is going to build a vintage C6 transmission designed for old-fashioned big-block power yet with engineering refinements that have come since the mid 1970s, including a 2.72:1 first and a 1.54:1 second gear. Although this gearing is engineered for the Modular V-8 and V-10 engines with their different torque curves, it works quite well in drag racing because vintage big-blocks offer you the acceleration advantage.

Seals

One of the most important elements of a transmission build is proper seal use and installation. Take note of what type of seal is on each component during disassembly to ensure the same type of seal goes back on. You may find transmission failure from time to time where seals were improperly installed or wrong seals were used. When you install clutch pistons, do it gently and carefully to prevent seal damage, which is one of the leading causes of transmission failure.

Forward Clutch

Because Brian is going with the TCI Low Gear, Low-Drag package here, there are significant changes happening to the geartrain you need to be aware of. The forward clutch hub, borrowed from the E4OD parts bin, is different than the C6's and arrives as a part of the TCI kit. Torrington bearings throughout greatly reduce frictional issues.

The TCI Auto six-pinion forward and rear planetary carriers provide greater load distribution qualities than you find on the older C6 three-pinion units, which makes our C6 better for durability in racing or towing.

Some transmission techs endorse removal of the wavy washer in the forward clutch for the addition of a clutch friction. We have five clutch frictions as it is, which is satisfactory, and plenty for the street.

And while I am on the subject of clutches, some transmission shops dress the steels for improved traction. However, I've never seen steels come out of a box dressed, but instead as raw steel. Clutch drums should be dressed with 400- to 600-grit paper because traction here is critical to better upshifts. The wider you can spec your band width, the better. I cannot overstress the importance of checking clutch clearances followed by an air-check on each.

During forward clutch assembly, pay close attention to the Belleville spring condition. Unseen hairline cracks can cause the Belleville spring to break during that first road test. I am often inclined to suggest a new Belleville spring with every transmission build because this piece gets stressed to where cyclic fatigue takes its toll.

Another issue to watch for is any clutch piston/drum irregularities or scoring. It's the minute things you cannot see (ragged edges and nicks) at first glance that can cause transmission failure; seals get damaged by these irregularities during installation or start-up.

Geartrain Endplay

C6 transmission geartrain endplay is the total movement of all your transmission's moving internal parts, which is adjusted at the number-1 selective-thrust washer at the front pump. Endplay should be .008 to .044 inch for the C6. Ford suggests these washers and thicknesses. If you're going racing, seasoned transmission builders suggest .024- to .030-inch endplay.

Selective-Thrust Washer Thicknesses

Thickness (inch)	*Identification Number*
.056–.058	1
.073–.075	2
.088–.090	3
.103–.105	4
.118–.120	5

Building the C6

As shown here, Brian from Tom's Transmissions builds a street/strip C6 transmission and shows you how to improve both durability and performance with components from TCI Automotive. Here's what's on the bench at Tom's Transmissions from TCI Automotive:

- 1966–1976 C6 Master Racing Overhaul Kit
- 1966–1976 C6 Low Gear, Low-Drag Planetary Set
- C6 Direct High-Performance Friction Clutch Steels (.075 inch)
- C6 Reverse High-Performance Friction Clutch Steels (.075 inch)
- C6 Powerband High-Performance Flex Band

Important!

1 Replace Shifter Seals

After you replace the manual and kickdown shifter seals this is the end result. Because the C6 has a history of popping into reverse gear without advanced notice, you want to double-check the detent and make sure it is solid and secure. This is a critical safety item; so pay very close attention to detail here.

2 Lubricate Seals

C6 build-up is in subassemblies beginning with the forward clutch shown here. Lubricate clutch-piston seals with assembly lube and roll them into place. Outer seal goes on the piston and inner seal goes inside the clutch drum.

Forward Clutch Assembly

1 Press In Forward Clutch Piston

Gently press into place the forward clutch piston, paying very close attention to seal activity (left). Use great care, walking each seal carefully into place using a seal manipulation tool (right). It is very easy to damage the seals.

2 Install Lock Ring

This lock ring secures forward clutch piston.

3 Insert Belleville spring

Forward clutch gets the Belleville clutch-piston return spring, also known as a disc spring. This spring returns the piston to rest when hydraulic pressure stops.

4 Install Pressure Plate

Forward clutch's forward pressure plate is next.

5 Install Wavy Washer

After the pressure plate, install this wavy washer. Some transmission professionals suggest deleting the wavy washer and adding a friction. Not everyone agrees on this one. Because there are already five clutch frictions here, this modification isn't necessary.

6 Insert Frictions and Steels

Alternate clutches and steels with this clutch friction being first after the wavy washer. This particular forward clutch gets five frictions and four steels. The close-up on the right shows how clutch frictions are internally driven. Clearances are determined by snap-ring thickness. There are three selective snap-ring thicknesses for setting forward clutch clearances: .065 to .069, .074 to .078, and .083 to .087 inch.

7 Install Pressure Plate

After you install all clutches and steels, install the pressure plate.

8 Inspect Forward Clutch

Completed forward clutch looks like this at rest with no pressure applied (left). With pressure applied, forward clutch looks like this (right) with clutches compressed and Belleville spring distorted. Note: number-3 and -4 thrusts have both been lubed and installed mid-clutch.

Reverse-High Clutch Assembly

1 Install Reverse-High Piston Seal

Reverse-high (direct) clutch piston receives its seal and generous amounts of lubrication. There's also an inner seal, which is installed inside the reverse-high clutch drum. Although transmission assembly lube is used here, you can also use Vaseline for seal lubrication.

2 Install Reverse-High Clutch Piston

Gently press the reverse-high clutch piston into place making sure there's no seal distortion or damage.

3 Install Return Springs and Retainer

Return springs and retainer are next. You can use a clutch spring compressor (shown) or C-clamps for spring compression and snap-ring installation. The number of springs depends on application.

Critical Inspection

4 Inspect Measure Reverse-High Clutch Pressure Plate

Inspect the reverse-high clutch pressure plate for heat checking and thickness. Thickness determines clutch clearances though adjustment is determined by selective snap-ring thickness.

5 Install Clutches & Steels

Install reverse-high clutches and steels next. This one gets five clutches.

6 Install Pressure Plate

Pressure plate is installed next, after clutches and steels. This is the number-2 thrust washer. Recommended clutch clearances are .022 to .036 inch. Selective snap-ring thicknesses to adjust clearance are .065 to .069, .074 to .078, and .083 to .087 inch.

7 Install Forward Clutch Hub

With TCI's C6 Low Gear Planetary set, some changes have to be made to accommodate the swap. On the left is a C6 forward clutch hub. On the right, a E4OD forward clutch hub, which enables us to use TCI's Low Gear Planetary gear set. Move thrust washer over the E4OD hub. This is one of few spots that doesn't have a Torrington bearing.

8 Insert Retaining Ring

Forward clutch gets a retaining ring for a secure union with the reverse-high (direct) clutch drum.

9 Mate Clutch Packs

Forward clutch is mated to the reverse-high clutch drum. This involves working the forward clutch back and forth until it completely seats in the reverse-high clutch drum. Both assemblies should be completely flush if properly seated.

10 Install Thrust Washers

Numbers-3 and -4 thrusts are installed. This is a low-friction Torrington bearing.

11 Install Hub

Carefully spline the forward clutch hub into the forward clutch. Pay close attention to how the hub feels and confirm it has seated.

Planetary Gearset Installation

1 Planetary Upgrade Option

Now to the real meat of this C6 build—TCI's Low Gear, Low-Drag Planetary set for 1966–1976 C6 transmissions. On the right is a C6 three-pinion forward planet carrier. On the left is TCI's six-pinion forward planet, which offers durability plus 2.72:1 first gear and 1.54:1 second gear for crisp acceleration. Low-friction Torrington bearings reduce internal friction by a wide margin.

2 Insert Thrust Bearing

Ready the TCI Low Gear forward planet for installation with this number-5 Torrington thrust bearing.

3 Install Forward Planet

TCI Low Gear forward planet slips right into the forward clutch hub ring gear.

4 Insert Bearing

TCI forward planetary gets this Torrington thrust bearing.

5 Install Shell & Sun Gear

Input shell and sun gear are next, splined into the forward planet. This subassembly is also unique to the TCI Low Gear Planetary gearset for the C6.

6 Inspect Rear Planet

TCI's Low Gear package (left) also includes a six-pinion rear planet for durability and improved gear ratio. On the right is a factory three-pinion example.

Low-Reverse Clutch Assembly

1 Install Roller Clutch

Low-reverse clutch hub gets its new TCI one-way roller clutch.

2 Install Clutch Piston Seals

Lubricate inner and outer seals before they're installed. Be careful with some applications, which have a lip seal that must be installed with the lip toward the pressure source. Install a lip seal backward, and you do not contain the pressure. Your first clue is soft or no engagement. Keep in mind what direction the pressure is coming from.

3 Install Clutch Piston

Carefully press the low-reverse clutch piston into the case, watching out for seal distortion. When secure, there are 24 clutch-piston return springs to install.

4 Insert Spring Retainer

Secure the low-reverse clutch-piston spring retainer with a spring compressor and snap-ring. This gets very tricky if you don't have a spring compressor. You have to rent or borrow one.

5 Install Inner Race

Next install the one-way roller clutch inner race. Make sure to lubricate all surfaces.

6 Install Hub

Install the low-reverse clutch hub, making sure roller clutch fits neatly and smoothly around the inner race. There should only be one-way rotation.

Critical Inspection

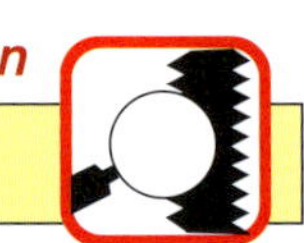

7 Inspect Steels

Inspect low-reverse clutch steels for heat checking.

8 Prepare To Install

Lay out low-reverse clutches, steels, and pressure plate across a workbench in the order of installation.

9 Install Steel

Low-reverse steel goes in first.

10 Install Friction, Steels, Clutches

Low-reverse clutch friction is next, followed by steels and clutches. These clutch frictions are splined into the low-reverse clutch hub.

11 Install Pressure Plate

Low-reverse clutch pressure plate goes in last, including the selective snap-ring, which is sized to set clutch clearances based on thickness.

Governor Installation

1 Inspect Governor Assembly

Disassemble and inspect the governor assembly. You want freedom of piston (valve) movement and proper spring pressures. Check for rattle. All passages should be clear.

2 Reassemble Governor Assembly

After you reassemble the governor assembly it is ready for installation. There is no gasket or seal, only perfectly machined surfaces. Carefully torque governor fasteners to 80 to 120 in-lbs. Never go on feel because overtightening means distortion and valve malfunction.

3 Inspect Rings

Check governor distributor sealing rings for fit and contact surfaces for scoring. Any scoring is unacceptable and must be machined out or replaced.

4 Install Governor

First lubricate the governor distributor sealing rings then slide the assembly onto the output shaft. Remember, there's a ball that works as a Woodruff key to the spline output shaft and governor. If you forget this ball, your C6's governor will not work.

5 Inspect Pawl

Inspect and assemble the parking pawl. This weak spot for C6 transmissions must be closely inspected. Check for solid, secure operation because this is a safety item.

6 Install Washer

Lube and install the number-10 thrust washer.

7 Install Parking Gear

Closely inspect parking gear teeth for any damage or potential failure points. Install the gear in the governor distributor sleeve.

8 Install Distributor Sleeve

Press the governor distributor sleeve into place. Torque the bolts 12 to 20 ft-lbs.

9 Install Output Shaft

Lubricate and install the output shaft, making sure it is firmly seated in the case. If the output shaft is hard to turn, check sealing rings, contact surfaces, and governor sleeve for any irregularities. Never force the shaft. Install the C-clip and slide it up to the roller clutch's inner race.

10 Install Bearing

This Torrington thrust bearing is next.

11 Inspect Installation

The installed low-reverse clutch package should look like this. The low-reverse clutch and hub, one-way roller clutch, and the output shaft are visible.

12 Install Ring & Hub

The reverse ring and hub slide into the low-reverse clutch hub. Make sure all surfaces are lubed.

13 Install Rear Planet

TCI six-pinion rear planet is next and should spline right into the reverse ring gear.

14 Insert Torrington Thrust

Number-7 Torrington thrust bearing is next. Torrington bearings reduce internal friction and free up power.

Geartrain Installation

1 Install Forward Geartrain

The forward geartrain splines right in like this. Input shell and sun gear spline into rear planet.

2 Install Intermediate Band

TCI Powerband intermediate band goes in next. This takes careful navigating to where the band-to-strut contact points fall in line with the adjustor and servo.

3 Align Intermediate Band

Intermediate band and both struts are in place. On the left is the servo apply lever and strut. On the right is the adjuster and strut. Make sure struts are firmly seated (refer to step 13 on page 110).

4 Install Intermediate Servo Apply Lever

C6 transmissions are fitted with a number of different intermediate-band servo apply levers. Each is identified by a letter code and fulcrum design. The base Ford part number is 7330. The A lever (left) has a 1.65:1 ratio. The K (right) is 1.575:1.

5 Inspect Input Shaft

Input shaft should be inspected for abnormal wear and twist. This one has corrosion and should be replaced, though it wouldn't be an issue in a non-performance application. TCI has a 31/30 spline hardened input shaft for C6 transmissions made from billet steel, a natural complement to the Low Gear system already installed.

6 Insert Splines

Input shaft splines into the forward clutch.

7 Use Grease For Pump Gasket Retention

Apply a thin film of transmission lube for gasket retention prior to front pump installation.

8 Install Front Pump Gears

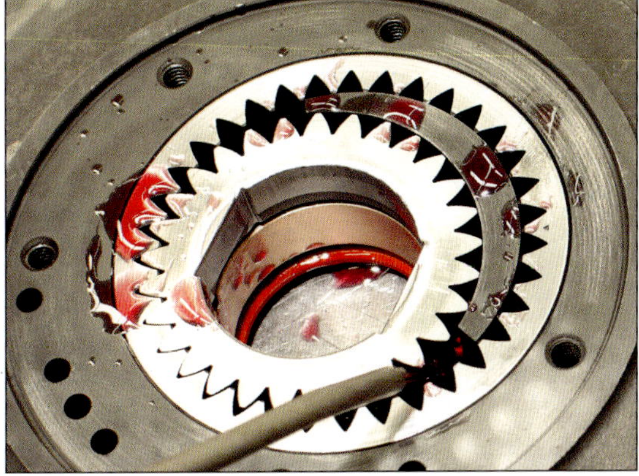

As long as there's no damage or excessive wear to the pump cavity, all you have to sweat is bushing and seal replacement. Inspect pump gears and cavity to be satisfied with their condition. Install the gears and fill the cavity with transmission fluid for a good wet start.

Final Assembly

1 Assemble Pump

Pull together the pump halves and torque the bolts to 28 to 40 ft-lbs.

2 Install Rings

Front pump sealing rings are gapless. Hook the ends and overlap them to be flush.

3 Install Pump O-Ring

Lubricate and install the pump-housing O-ring seal, making sure seal is square with groove.

4 Install Thrust Washer

This is the number-1 selective thrust washer, located at the front pump, which controls geartrain endplay. If endplay requires adjustment, there are a variety of thicknesses available (.008 to .044 inch).

5 Air-Check Clutch

Reverse-high (direct) clutch can be air-checked through the front pump.

6 Install Pump Gasket

Position the front pump gasket and check for proper alignment.

7 Install Pump Assembly

Line up and install the front pump assembly. Use a center punch for alignment. Install the bolts and use them to run the pump down. Torque is 28-40 ft-lbs. Monitor seal status as the pump seats.

8 Choose Intermediate Servo

This 2.465-inch-diameter servo from TCI (left) offers 25- to 35-percent more apply area than the Ford R servo. This means more apply power for your C6's intermediate band. This billet-aluminum intermediate assembly (PN 423005) is a big step up from the Ford R servo (right), which is typically recommended for street-performance C6s.

9 Install Intermediate Servo Piston Seals

Lubricate and install the seals on the TCI intermediate band servo piston.

10 Install O-Ring & Gasket

TCI servo cover receives its O-ring seal along with a gasket that goes between the cover and main case.

11 Assemble Cover & Piston

Servo cover and piston are mated and ready for installation.

12 Install Cover & Piston

Install the TCI servo cover and piston, taking extra care to not damage seals going in. Get cover and piston square with the case and gently press into place. Torque is 12 to 20 ft-lbs.

13 Adjust Intermediate Band

Torque the intermediate band adjustment with 10 ft-lbs, then back off 1½ turns. Hold the adjustment stud and tighten the locknut 35 to 40 ft-lbs.

14 Air-Check Clutches & Servos

All transmissions built by Tom's Transmissions get an air-check that includes clutches and band servos.

15 Install Tailshaft Sleeve

The tailshaft housing gets a new TCI sleeve. Pay close attention to the sleeve drain-hole location, which must line up with the casting provision.

16 Install Valve Body

With close attention to manual shift and kickdown levers, install the valve body.

17 Check Valve Body Function

Although the governor has already been installed, here is how to check valve body function through the governor passage at the tailshaft.

External Assembly

1 Inspect Throttle Valve

The throttle valve, which is actuated by the vacuum-operated throttle valve assembly (vacuum modulator). Check for a clean fit and lubricate with transmission fluid before installation.

2 Install Throttle Valve

Next install the throttle valve assembly and pin. This is a press-in throttle valve with bracket common from 1972-on. (Prior to 1972, a screw-in throttle valve and gasket is used.) Pin lengths tend to vary depending upon application and what you want the transmission to do.

3 Inspect Pan Gaskets

If you're tired of leaking pan gaskets, pay close attention to pan condition. Warped and distorted pans leak regardless of what you do for a gasket. Pan contact surfaces must be true and void of distortion. If the pan rail cannot be dollied true, replace the pan. Cast aluminum pans offer fluid capacity and cooling capacity. If you must use a steel pan, find one with perfect rails and use a thick, high-quality gasket. You could use a super-thin film of Form-A-Gasket non-hardening sealer on leakers but not everyone agrees with this approach.

4 Transmission Fluid

You don't have to use Type F in old Ford automatics. It is more old-school with dated friction enhancers. For old Ford automatics today you can use Dexron III or Mercon IV, which are new technology. Torque converters must always receive 1 to 2 quarts for a good pump prime and quick lubrication on start-up.

5 Install Torque Converter

If installed incorrectly, the torque converter can cause pump and converter damage. Carefully work the converter, stator, and input shaft until the converter seats. If you can get your hand between the converter and bellhousing, the converter isn't properly seated.

Torque Converter Stall Speeds

Engine	*RPM*
351W	1,500–1,700
351C	1,500–1,700
390-2V FE	1,700–1,900
390-4V FE	1,750–1,950
390-4V FE GT	1,700–1,900
427-4V FE	1,850–2,050
428-4V FE	1,850–2,050
428-4V FE CJ	1,830–2,030
429-2V	1,820–2,020
429-4V	1,820–2,020

C6 Clearances

Function	*Measurement (inch)*
Geartrain Endplay	.008–.044 using selective-thrust washers
Turbine/Stator Endplay	.021 new
	.030 used
Intermediate Band Adjustment	Torque adjustment to 10 in-lbs, then back off 1½ turns; tighten locknut to 35–45 ft-lbs
Selective Snap-Ring Thicknesses (Forward Clutch)	.056–.060
	.065–.069
	.074–.078
	.083–.087
	.092–.096
Forward Clutch Clearances	.031–.044 at snap-ring
Reverse Clutch Clearances	.022–.036 depending on unit
	.027–.043 depending on unit

TCI-equipped C6 with the improved Low Gear package is ready for action. With the low gear ratio, acceleration is dramatically improved, as is the fierce reliability of E4OD internals and low-friction benefits of Torrington bearings instead of thrust washers.

Performance Tip

6 Install Low-Friction Torrington Bearings

One path to a better C6 is to install low-friction Torrington bearings between geartrain stages. Low drag equals better performance at the tailshaft. Torrington bearings handle friction better than those old thrust washers.

Torque Specifications

Location	*Foot-Pounds*	*Location*	*Inch-Pounds*
Torque Converter to Flexplate	20–30	Valve Body End Plate	25–30
Bellhousing to Case	40–50	Inner Downshift Lever Stop	20–30
Pump to Case	12–20	Reinforcement Plate to Valve Body	20–30
Roller Clutch Race to Case	18–25	Filter and Valve Body Half Bolts	50–60
Oil Pan	12–16	Valve Body to Main Case	95–125
Stator Support to Pump	12–16	Governor Body	80–120
Intermediate Servo Cover	10–14	Oil Tube Connector	80–120
Diaphragm to Case	15–23		
Governor Distributor to Case	12–16		
Manual Lever to Shaft Nut	35–45		
Tailshaft Housing	25–40		

C6 Shift Kits

Automatic transmission function, in theory, is simple. In practice, it is quite complex and has been worked out by powertrain engineers determined to give you smooth and reliable operation for at least 100,000 miles. To get there, you have to have a hydraulic control system that gives you all those things you want and expect from a C6 automatic transmission. You want a firm upshift, yet you also want a downshift during coast and deceleration that cannot be felt. And when you punch the accelerator, you want a quick and positive downshift that delivers rapid acceleration without delay. That's a lot to expect from one of the greatest feats of mechanical engineering in automotive history. And it has only improved in the years since those old iron automatics were new.

When companies like TCI Automotive develop shift modification kits like the Trans-Scat valve body kit, a lot of thought, research, and development has to go into that kit. And this is something you don't take lightly because of what you want from an automatic transmission.

TCI Automotive offers two Trans-Scat options for the Ford C6: Towing or Competition. In both cases, the Trans-Scat kit gives you a firm upshift, which is nothing more than solid clutch-and-band engagement to eliminate slippage. Stock automatic transmissions deliver a soft shift because most people generally don't want to feel the upshift. To get a soft shift, there has to be clutch and band slippage, which means slippage is programmed into valve body design.

Slippage comes from the gradual engagement of clutches and bands for a soft shift, which causes heat and the release of friction material into the transmission fluid. This is what drivers generally want from their automatics—a soft, unnoticeable shift, which is a product of slippage. Friction material in the fluid from slippage damages seals, which hinders line pressure and causes additional slippage and transmission failure.

There's more to the C6 Trans-Scat kit than just performance. It is all about giving your C6 longevity because this kit gets rid of destructive slippage. A firm upshift means no slippage, better performance, and longer transmission life.

Installing a TCI Trans-Scat

So what happens when you install a shift modification kit? Without becoming too technical here, a shift modification kit changes line pressure and timing to produce a firm upshift, mostly at wide-open

Trans-Scat from TCI

The Trans-Scat valve body kit (PN 360000) for C6 transmissions offers a nice improvement in shift quality and durability. There are two basic options: High-Performance Street/Strip or Severe Duty (for towing and recreational vehicles). Best of all, you get to keep all Drive-1 functions regardless of what package you choose. Go with the High-Performance Street/Strip version and you get a good snap to your C6's upshift–a nice, crisp upshift, which saves unnecessary wear and tear on clutches and bands. It enables your C6 to deliver horsepower and torque at wide-open throttle without slippage and power loss for tire-barking performance. Severe Duty kit upshifts aren't as harsh as High-Performance Street/Strip, but are still firm enough to delivery solid durability under the toughest driving conditions.

The Trans-Scat kit includes:

- Drill bit, 1/8 inch
- Servo modulator valve plug
- Check ball, 1/2 inch
- Blue pressure-regulator spring (1967–1974)
- Orange pressure-regulator spring (1975-on)
- Violet 2-1 scheduling-valve spring
- Plain or natural-metal-color accumulator spring
- Pan gasket
- Valve body separator plate
- Filter and gasket

You are going to need these tools:

- 3/8-inch-drive ratchet or speed handle
- 1/2-, 3/8-, and 5/16-inch sockets
- 13/16-inch wrench
- 0 to 250 in-lbs torque wrench
- 0 to 250 ft-lbs torque wrench
- 1/4-inch drill
- small flat-blade screwdriver
- small file or deburring tool

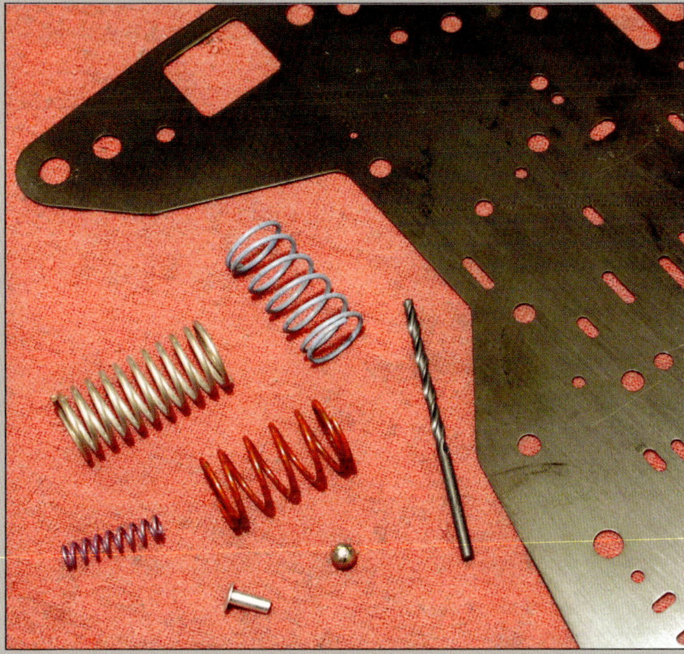

The TCI Automotive Trans-Scat valve body kit (PN 360000) is simple in scope. Where it gets complicated is modifications that have to be made to the valve body that require drilling. You must pay very close attention to detail here because once you drill, there's no turning back. If you drill wrong, you have scrap or a potentially serious transmission malfunction.

throttle. It also delivers a firmer shift under normal driving conditions. Instead of gradual applications of hydraulic pressure to clutches and band servos, pressure arrives more quickly and firmly, which prevents slippage. It is not only about increased line pressure, but the all-important timing.

The C6 valve body as it appeared in the mid 1970s, with external detent. In this chapter, Tom's Transmissions shows how to knock this valve body apart, clean it, and install the Trans-Scat valve body kit from TCI Automotive, for firm upshifts, good street performance, and longevity.

1 Remove Filter

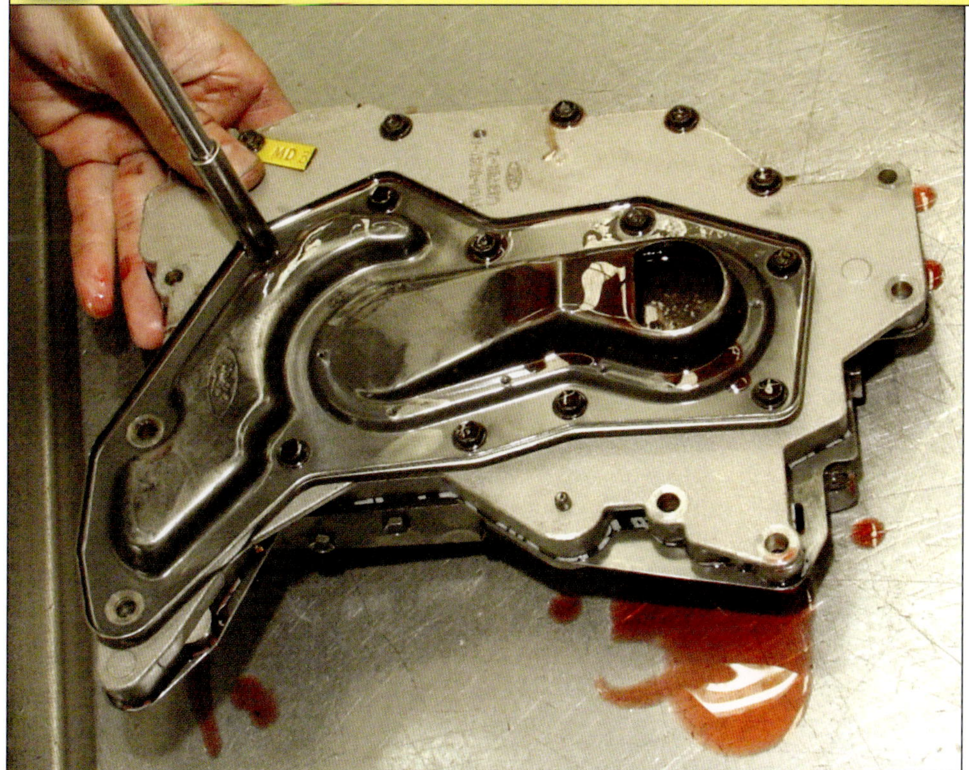

Valve body disassembly begins with filter removal and disposal. You're going to need 3/8- and 5/16-inch sockets.

Documentation Required

2 Remove Valve Body Bolts

C6 valve body consists of two halves. Remove remaining bolts with a 3/8-inch socket. Before you separate the halves, remember to inspect the check balls and two spring-loaded valve assemblies. As you carefully separate the halves, take note of check ball and valve locations.

3 Note Position

You want the larger upper half on top when the halves come apart.

4 Remove Relief Valves & Springs

Both the drain-back valve and converter pressure-relief valve look like this and are located in the valve body. Carefully remove and reinstall the springs and valve discs in the same locations.

5 Remove Separator Plate

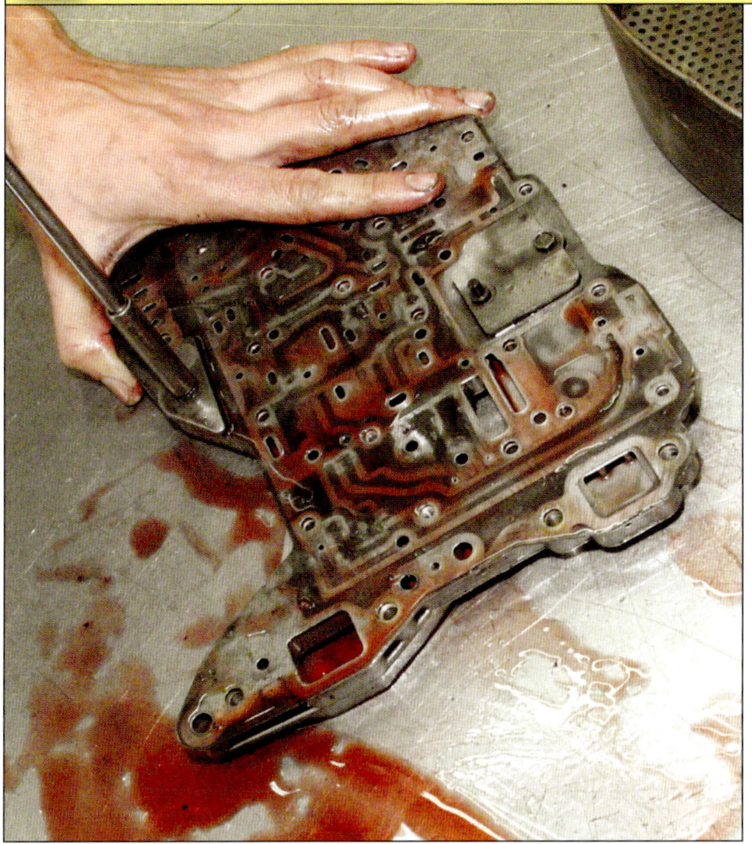

Next remove the separator plate from the upper half of the valve body.

Critical Inspection

6 Examine Valve Body

When you closely examine the C6 valve body's lower half, it appears very complex. Actually, each passage and valve performs a specific function, channeling line pressure to clutches and the intermediate band servo. Although check balls and valves are not shown here, you need to know where they're located.

7 Wash Valve Body

Wash the valve body using a petroleum-based solvent. Never use a water-based solvent. Water aggravates corrosion between steel and aluminum parts.

8 Note Pressure Boost Valve

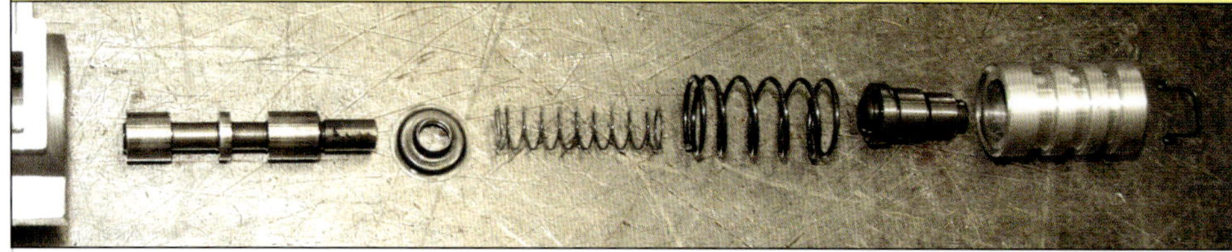

Trans-Scat installation begins at the pressure boost valve. This is the stock configuration. Early C6 valve bodies have a retainer plate with two screws removed with a 5/16-inch socket. Later versions don't have this plate, but instead a retainer clip that holds the pressure boost valve, sleeve, springs, and main pressure regulator valve. In any case, be prepared for a spring-loaded assembly where parts can be lost on release.

9 Use Correct Spring

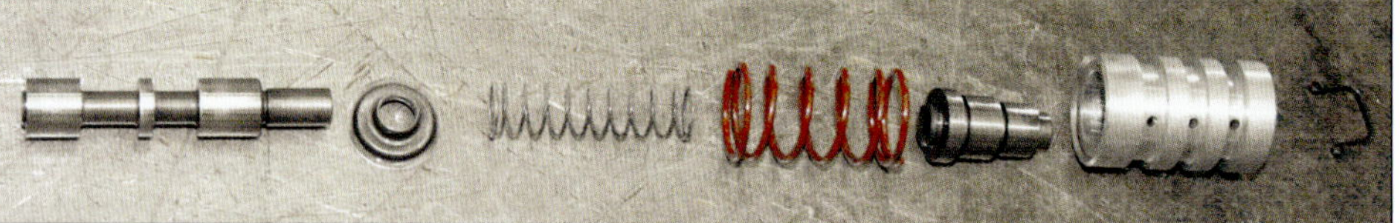

This application calls for the 1975-on orange pressure-regulator spring with no other changes. The orange spring has more tension than the blue one.

10 Install Pressure Boost Valve

The entire assembly goes back in and ends with the sleeve and pressure boost valve shown here. After this valve package is flush, reinstall the retaining clip or plate depending on model year. Always check for freedom of valve movement before continuing.

11 Remove Long End Plate

Using a 5/16-inch socket, remove the long end plate. There are seven different valves located here and they demand your close attention to detail.

12 Install Servo Modulator Valve

Next is the intermediate servo modulator valve. The factory installed a spring here to modulate servo pressure.

13 Replace Spring With Plug

TCI eliminates this spring and goes with a plug, which cannot stick out of the valve body. If it does, it must be ground down to where it is flush or below deck. If you must grind the plug, clean and deburr it before installation.

14 Remove Servo Accumulator Valvespring

Next in line is the intermediate servo accumulator valve. This can be confusing because the intermediate servo accumulator valve is fitted with one spring early in production and two later on. This 1975 C6 valve body has one spring, which adds to the confusion. Replace this spring with the natural-metal spring from the kit. There's less spring tension from this TCI spring.

15 Make Modifications (If Applicable)

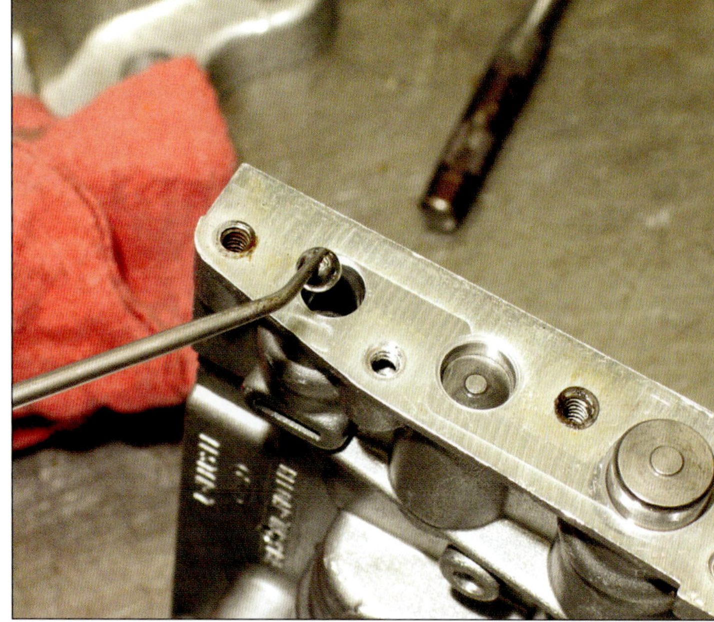

Changes need to be made to the cut-back valve if you're building for high-performance street/strip. Remove cut-back valve in this bore and replace with a 1/4-inch check ball. Then reinstall cut-back valve. For severe duty/towing, no modifications are required here.

What Do All These Valves Do?

From left to right:

- Intermediate Servo Modulator Valve
- Intermediate Servo Accumulator Valve and Spring
- 2-3 Back-Out Valve, 3-2 Shift Timing Valve
- 2-3 Shift Valve, Throttle Modulator Valve
- 1-2 Shift Valve, DR2 Shift
- Line Pressure Coasting Regulator Valve
- Cut-Back Valve ■

This view with the long end plate removed shows valve arrangement from left to right.

Valve Body Reassembly

1 Reinstall Long Plate

Reinstall the long-end plate, using a 5/16-inch socket. Torque is 20 to 30 in-lbs. After installed, make sure all valves have freedom of movement.

2 Remove Short End Plate

Remove this smaller end plate next, using a 5/16-inch socket to expose throttle boost and manual-low 2-1 scheduling valves.

3 Remove Valve & Spring

The throttle boost and manual-low 2-1 scheduling valves are both shown here. Throttle boost valve is installed in the valve body. Remove only the manual-low 2-1 scheduling valve and spring.

4 Replace Spring

Swap out the manual-low 2-1 scheduling valve spring with the violet spring from your TCI kit. Reinstall the manual-low 2-1 scheduling valve and spring and the end plate. Torque is 20 to 30 in-lbs.

5 Note Proper Drilling Locations

TCI instructions include a template, which shows you where to drill, using the 1/8-inch bit included in the kit.

6 Drill in Appropriate Locations

Using your C6's old separator plate, compare it with the illustration in your TCI instructions One of these holes lines up with the hole already in your separator plate. Mark the existing hole. Take the TCI separator plate in your Trans-Scat kit and mark the position of the hole from your existing separator plate. Take the 1/8-inch drill bit and drill the new hole in your new TCI separator plate. Discard the old plate. Deburr the hole with a larger bit.

7 Reassemble Upper Half

Reassembley the upper half of your C6 valve body with new gaskets from the TCI kit. Bolt torque is 20 to 30 in-lbs.

8 Install Check Balls

Check ball installation is next. Look for round pockets to ascertain location. This is the 1/4-inch-diameter throttle-pressure relief ball and spring.

9 Install Disc & Spring

The torque-converter pressure-relief valve disc and spring. Nearby is the 2-3-shift check valve and spring, which look like this valve—you don't want to get the two mixed up.

10 Assemble Valve Body

Two valve body halves go back together. Torque is 20 to 30 in-lbs. Valve body is ready for installation. Don't forget the filter.

TECH TIP

Valve Bodies

There are three significant C6 valve body changes: 1966 (a standalone piece for the Dual-Range C6 Cruise-O-Matic), 1967–1976, and 1977–1996. Check your valve body casting numbers (both halves) to ascertain which you have. Also remember C6 valve bodies went from internal detent to external detent with a case-mounted roller in the mid 1970s. Make sure you have a good case and valve body match. ■

CHAPTER 8

Torque Converters

Torque converters are probably the most misunderstood component in an automatic transmission, yet they're the simplest in both theory and function. Think of a torque converter like a water wheel in an old saw mill: the waterwheel is driven by fluid in motion. A torque converter works on the same principle—a fluid coupling or clutch that slips when the vehicle is stopped and transfers power as engine RPM increases and gets fluid moving. A torque converter, by its very nature as a fluid coupling, also dampens engine combustion pulses to achieve smoother operation.

A Bit of History

The use of torque converters dates to the early 1900s. The Germans were among the first to use torque converters in automobiles, trains, and industrial machinery. The first US automaker to use a torque converter was Chrysler in the 1939 Imperial, known as Fluid Drive. General Motors followed that act in the 1940 Oldsmobile. Ford then followed suit in 1942 with a BorgWarner derivative in Lincoln and Mercury automobiles.

These early uses of torque converters didn't work very well on start-out because there was no torque multiplication in those days. In fact, torque converters were called "fluid couplings" at the time because they didn't multiply torque. General Motors was first with a true torque converter in the 1949 Buick Dynaflow transmission. Ford followed GM's lead in 1950 with the first Ford automatic designed and manufactured by BorgWarner. GM's legendary Powerglide 2-speed automatic came along in the mid 1950s and became a favorite with drag racers as time went on.

Torque Converter Function

Thanks to the basic principles of hydraulics, a torque converter puts fluid in motion to do our work. Fluid is thrust into motion to drive components in a process known as hydraulics. The same principle that stops your car in the braking system or operates your power steering is what gets it going in an automatic transmission. And if everything is working properly, the work is done smoothly and efficiently.

A torque converter consists of four main components:

- Impeller, which is tied to the crankshaft and puts fluid in motion
- Stator, which directs fluid under pressure to the turbine
- Turbine, which is tied to the transmission input shaft, driven by fluid in motion from the impeller and stator
- Cover or shell, which is welded to the impeller

The cover/shell and impeller are welded together to form the main torque converter shell, which drives the transmission's front pump to provide hydraulic pressure for operation and lubrication. The impeller drives fluid through the stator to the turbine, which is tied to the transmission's input shaft. As engine speed increases, fluid flow is directed through the stator to the turbine, which drives the turbine and transmission input shaft to get you moving.

> **TECH TIP**
>
> **Stall Speed**
>
> Choose your torque converter stall speed based on how you intend to drive most often. A high-stall converter will drive you crazy on the street. Choose a happy medium between street and weekend racing if you intend to do both. A 2,400-rpm stall is a nice compromise for street and strip. ■

Stall Speed

The point at which the impeller begins to drive the turbine is known as stall speed. Most stock torque converters "stall" at around 1,500 to 1,900 rpm of engine speed. High-performance torque converters stall at higher engine speeds because you want the engine well into its power band when the converter stalls (begins to move the turbine and vehicle). For example, a 2,400-rpm-stall torque converter doesn't begin to move the vehicle until engine speed reaches 2,400 rpm. The same can be said for a racing converter with a stall speed of 3,600 rpm. You want the engine making power when it hooks up (stalls) with the transmission's input shaft.

Stator and Clutch

Stall speed is determined mostly by stator design. The stator is the "brains" of a torque converter because it manages fluid flow from the impeller to the turbine. This is what makes a torque converter a torque multiplier. The engine's torque output is multiplied at least twice over, thanks to the stator. Most torque converters multiply torque in a 2.5:1 ratio over actual engine torque at stall speed. Within the stator is the one-way clutch splined onto the transmission's stator support shaft. The one-way clutch allows the stator to rotate in one direction only with the engine's crankshaft and converter impeller/shell. Torque conversion or multiplication happens at stall speed with the stator stationary before the turbine begins to move. When the turbine gets underway with the vehicle in motion, the stator moves at the same speed as the turbine.

You can actually feel this process happen as you step on the gas and feel the vehicle accelerate. During hard acceleration, you can feel torque multiplication (stator stationary or

There's no magic behind torque converters. Open one like this one from TCI Automotive and you can see it's basic fluid dynamics and propulsion. After-market performance converters are all about higher stall speeds and furnace-brazed construction that can take a beating.

This is how the torque converter interacts with your C4 or C6 transmission. Fluid under pressure through the stator drives the turbine and transmission input shaft. Stator support carries the torque converter and is also an integral part of the transmission's front pump.

The impeller is basically an engine-driven pump that moves fluid to and through the stator to the drive turbine. As long as the impeller receives a continuous supply of fluid, it continues to power the turbine.

Impeller generates fluid flow, which travels through the stator to drive the turbine. Converter's outer shell, driven by the engine's crankshaft, drives the transmission's front pump. Front pump operates only with the engine running.

Fluid flows aggressively through this stator, multiplying your engine's torque. When the stator is turning more slowly than the impeller, you get torque multiplication. As the stator catches up with vehicle speed, torque multiplication stops.

Although torque converters tend to look the same, what they do can be very different, especially when it comes to stall speed and acceleration. Locking torque converters, which are not used on the C4 and C6, have a built-in hydraulic clutch that contacts the shell for direct lock-up.

slower than turbine speed). As the vehicle gets up to speed, the stator slowly begins to rotate to crankshaft speed. Lean on the gas and stator speed falls behind and torque multiplication comes into play, which is when you feel gut acceleration.

Fluid Flow

There are two basic types of flow: rotary (circular) and vortex (roundy-round circular). When impeller and turbine speed are uniform, you have rotary flow in a circle around the converter's circumference. If there's a difference in impeller and turbine speed, flow becomes more vortex (tornadic) in nature.

As said earlier, the stator is what helps the impeller and turbine multiply torque. During acceleration,

Transmission Rebuilding Company (TRC) rebuilds its own torque converters with the latest technology and a strong eye on quality. With this shell cut open, you can see the torque converter's internals.

This is the torque converter's impeller, which propels fluid under pressure to drive the turbine and transmission input shaft.

The stator directs fluid under pressure to the turbine. Think of the stator as a fluid manager, which multiplies torque as it directs fluid into the turbine.

the stator turns at a slower speed than the impeller and turbine, which directs fluid flow more aggressively against the turbine blades. As the vehicle speed catches up with turbine speed, the impeller, stator, and turbine are all whirling around at the same speed. Any time you step on the gas, stator speed slows momentarily to help direct fluid and multiply torque.

The stator's one-way roller clutch.

Choosing a Torque Converter

Most manufacturers categorize torque converters by size and stall speed. Performance Automatic, for example, makes it easy for you to choose a torque converter for your street or race application because, on its website, it explains the differences. As the diameter of a torque converter decreases, stall speed goes higher, which is why race converters are generally smaller than street converters.

It is a good idea to discuss your performance needs and expectations with a sales/tech professional before ordering a torque converter. Transmission parts supply houses generally sell stock torque converters with 1,500- to 1,900-rpm stall speeds. These converters are off-the-shelf dead-stock pieces that are not always designed and constructed for performance purposes.

If you're seeking performance, it is wise to deal with aftermarket

The torque converter's drive turbine, which is splined to the transmission's input shaft.

performance transmission companies like Performance Automatic, B&M, and TCI Automotive, whose products are all available from Summit Racing Equipment.

Aftermarket high-performance torque converters are designed and constructed to take additional punishment, with features such as:

- Furnace-brazed fins for solid integrity (stock fins are slotted in place, but not brazed)
- Dynamic balancing for high-RPM use
- Needle bearings instead of thrust washers
- Heavy-duty stator and sprag/one-way clutch
- 400- to 600-rpm-over-stock stall speed

Converter Diameter and Stall Speed

Stock torque converters come in sizes around 11 to 13 inches in diameter with stall speeds around 1,500 to 1,900 rpm. This RPM range is where you want a street engine to begin applying torque. When you slip the transmission into gear, a stock converter provides a gentle nudge as engine torque is applied to the transmission's input shaft and forward clutch. When you have a higher stall speed, that nudge doesn't happen until the engine is closer to stall speed.

You want a higher stall speed on a street engine when the application of power is expected to be in the 2,400- to 2,600-rpm range. Weekend racers like having a high-stall torque converter that takes hold in this range because that's where the power is.

For example, if you have a hot cam and an aggressive induction system along with a rough idle around 1,000 to 1,200 rpm, you want a higher stall speed for better traffic light idle, higher in-gear quality, and proper application of power as RPM increases. You want the torque converter to take hold (stall) at 2,400 to 2,600 rpm as the engine begins to make power. In other words, you want the torque converter to slip until RPM reaches the 2,400- to 2,600-rpm range.

Intended Use

The type of torque converter you choose depends upon how you intend to drive the vehicle. Street cruisers do not need high-performance, high-stall torque converters. They don't even need a high-performance converter with all of the features mentioned above. If you're going racing on Saturday night, you probably need a higher stall speed to get your engine into its powerband for a blistering holeshot and solid hook-up off the line.

Stock engines normally make peak torque around 2,000 to 3,000 rpm, with peak horsepower coming in around 5,500 rpm. High-performance engines normally make peak torque around 3,500 rpm, with peak horsepower rolling in around 6,000 to 6,500 rpm. A stall speed of 1,500 to 1,900 rpm is perfect for street use with a mild engine because you want the converter to take hold at the beginning of the engine's off-idle rise in power.

High-performance engines begin to make power at a higher RPM, which is where you want a torque converter to take hold with a higher stall speed. If you run a high-stall converter with a stock engine, slippage occurs until your engine reaches the high stall speed. This makes normal driving difficult. This means your engine revs and doesn't begin to transfer power until the higher stall speed is reached.

Slippage and high stall speeds affect upshifts. At 5,200 rpm, the engine speed drops by 3,500 rpm with each upshift. If the converter

Older torque converters have drain plugs for servicing, needed every 30,000 miles or 3 years. Never completely drain the torque converter due to the risk of pump cavitation. Watch out for drain plug alignment with your Ford's flexplate. It must line up with a corresponding hole in the flexplate or you wind up distorting the flexplate.

isn't fully stalled at that point, you lose performance, which is wasted via slippage. This costs you precious time on the quarter-mile or at the traffic light.

Converter Efficiency

Torque converter performance isn't just about stall speed; it's also about how firm a converter hooks up when it does stall. This is known as a tight or loose converter. Torque converter manufacturers like B&M, TCI Automotive, and Performance Automatic employ techniques that make torque converters more efficient with less slippage. Much of the general technology is rooted in fluid dynamics and how fluid behaves under given conditions. The greatest factor in converter construction is stator design, meaning blade/fin shape and angle, which determines stall speed and slippage. And this fact alone helps determine your quarter-mile times and the way your Ford behaves on the open road.

TCR pressure-tests every torque converter it rebuilds.

Torque Converter Efficiency

Here's a simple formula for determining torque converter efficiency.

RPM = (MPH x Gear Ratio x 336) ÷ Tire Diameter

Let's use a hypothetical vehicle as an example. We're going drag racing and plan to cross the traps at 117 mph and 5,100 rpm. This vehicle has a 3.42:1 rear-axle gear ratio with tires that are 27 inches tall.That means:

(117 x 3.42 x 336) ÷ 27 inches = 4,979 rpm

This works out to 5,100 versus 4,979 rpm–a difference of about 200 rpm or 2.4 percent. This means you have an efficient torque converter because slippage is minimal.

In torque converter function, there is load and flash stall, which happen at different RPM ranges. Load stall, also known as foot-brake stall or power braking, happens at wide-open throttle with your foot squarely on the brake. When you hold the brake pedal hard and floor the accelerator, you can observe RPM as the engine overpowers your brakes.

Flash stall happens when you take your foot off the brake at wide-open throttle and the vehicle begins to accelerate. This is known as "flashing" the converter–engine power begins to go to work accelerating the vehicle. Flash stall normally happens around 1,000 to 2,500 rpm with a street converter and higher with a racing converter. Flash stall is the true stall speed of your torque converter.

Although manufacturers have published torque converter stall speeds in their catalogs and websites, the true flash stall speed of a converter cannot be determined until you try it out. Many elements determine flash stall speed, especially in a performance application, such as engine RPM, vehicle weight, gearing, tire size, and converter design. No two torque converters operate the same way even if they have the same part number and published stall speed.

CHAPTER 9

Removal and Installation

Although I focus great attention on how a transmission is built, installation and proper adjustment after the unit is installed are just as important to durability and longevity. To get longevity and durability, a transmission must be properly installed and adjusted, which means a methodical approach to getting a fresh transmission into service.

Let's begin the installation discussion with a few removal points; that'll get you off to a good start.

Removal

Transmission removal is on par with transmission teardown (see Chapter 3) because it is a forensics study in why the transmission may have failed to begin with. Ask yourself, "How was the transmission installed?" "Was everything in proper adjustment?" How was the transmission functioning when it was time for removal?"

Installation

Transmission installation requires great levels of care.

Torque Converter

Torque converter installation, although it sounds simple, is an easy step to mess up. It's mostly about feel because you cannot see the components to ascertain fit. Not only must the converter slide onto the input shaft and pump stator, it must also fit squarely into the pump and engage the drive gear. If it doesn't seat properly, you risk severe pump and torque converter damage before even getting started. Believe me, a lot of us have made this mistake—don't you make it.

As you are seating the converter, rock it on the stator and feel for proper input shaft, pump, and stator engagement. Then do the hand-fit check. If your hand can fit between converter and bellhousing, it is not seated. Feel for a solid bottoming out and do

If the transmission failed, removal is a part of learning how and why failure occurred. Sometimes failure isn't due to transmission problems, but instead due to poor initial installation and adjustment.

This is an opportunity to check out a transmission's condition before teardown. What leaks and why? Are the kickdown and shifter linkages properly adjusted? Does the throttle valve (vacuum modulator) function properly? What condition is the transmission mount in? Is there a transmission cooler? What about driveshaft and slip yoke? What color is the transmission fluid?

Torque Converter Installation

Use a torque converter holder for removal and installation to prevent the torque converter from falling out and to keep it seated. This is especially important during installation, when you want the torque converter to remain seated. ■

You can fabricate a simple torque converter holder from 2-inch band iron and drilling a 1/2-inch bolt hole. This keeps the torque converter in place during removal and installation.

the hand-fit check. Then rotate the torque converter and see how it feels. If there's binding or noise (grinding), it is not properly seated. Again, perform that hand-fit check between converter and bellhousing.

After the converter is seated, use a fabricated retaining device, even if it's a simple piece of band iron, to keep it in place.

Cooling System

Before installation, check out your transmission's support system. The cooler lines and cooler must be flushed thoroughly to remove any particulates caused by transmission damage. Tom's Transmissions uses a system called Hot Flush HF345S, which pressure-flushes transmission coolers and lines with a quick-reversing hot-fluid action that slams fluid back and forth to dislodge any debris that can harm a fresh transmission. Particulates are trapped in super-fine filters that ensure damaging debris won't harm your new transmission.

What shape are your transmission cooler lines in? I've seen so many cars with patchwork transmission cooler lines connected together in pieces—

The Hot Flush system from Tom's Transmissions is an outstanding transmission purification system (left). Using a hot-flush, high-pressure surge system, it removes all impurities, which are trapped in a filter (right).

easy spots for destructive debris to accumulate. I strongly recommend against using hose between transmission and cooler. Hard-line your system from transmission to cooler, using a minimal amount of hose. Under ideal circumstances, you hard-line 100 percent of your transmission's cooler lines.

Linkages

Inspect manual shift and kickdown linkages for integrity and proper adjustment. Remember, transmission failure isn't always due to internal problems; there can be external ones like improper kickdown and manual linkage adjustments. Improper manual linkage adjustment is an easy mistake to make because a lot happens between your hand and the manual shift valve down under, especially with a column shifter.

Manual shifter adjustment needs to be on target per your Ford Shop Manual. Whether it is a console or column shift, Ford gives you plenty of room to adjust. Your adjustment should be in the middle of Ford's adjustment range.

Vacuum Modulator

And while I am on the subject of adjustment, vacuum modulator and control pressure should be a part of all this. Using a pressure gauge and following Ford's numbers, adjust the vacuum modulator. Most vacuum modulators are already properly adjusted. But you want to be sure with a new transmission. Never leave this one to chance.

Also make sure you have a working vacuum modulator and a reliable vacuum source. An engine with a lumpy camshaft and 14 inches of vacuum isn't going to deliver a reliable vacuum signal, which causes the transmission to shift erratically. So if you're having transmission problems with a rough-running high-performance engine, check your vacuum signal first before blaming the transmission.

Inspection

While your transmission is out, it is time to inspect things like your engine's oil pan gasket and rear main seal for leaks. And if you're going to take care of these items, now is the time to do it.

What about your starter? What is the starter drive's condition?

What about your flexplate? The ring gear should be free of damage and runout.

What about the block plate? Is it the right one for your application?

While your driveshaft is removed, inspect the slip yoke for scoring and abnormal wear.

What about universal joints? This is the time to se rvice your driveshaft and make it serviceable.

Next is the transmission crossmember and mount. A fresh mount keeps things secure around the driveshaft's centerline.

And finally, what about your speedometer cable and drive gear? This is the time to inspect and make sure you have the correct drive gear and make sure your speedometer cable is in top condition.

Flexplate Installation

1 Inspect Flexplate

First inspect the flexplate, checking for runout and ring gear damage. If all is true to mark and there are no cracks around bolt holes, the flexplate may be reinstalled.

2 Install Reinforcement Plate

Ford calls this the flywheel reinforcement plate (PN C2OZ-6A366-A), which must be installed with the flexplate. And like the flexplate, it installs only one way.

Critical Inspection

3 Check for Leakage

While your transmission is out, check for rear main seal and oil pan leakage. Now is the time to correct any engine problems.

4 Line Up Correctly

When you're preparing flexplate for installation, get all of the holes lined up in both flexplate and reinforcing plate because they only install one way. Use a reference mark to ease installing the reinforcing plate, flexplate, and crankshaft.

5 Use Thread Locker

Always use a high-temperature thread locker on crankshaft bolt threads to both secure and seal because these bolts thread directly into the crankcase.

6 Torque Bolts

Torque bolts crisscross to 75 to 85 ft-lbs in thirds: first, to 28 ft-lbs, then to 56 ft-lbs, and finally to 75 to 85 ft-lbs. Don't forget thread locker.

7 Check Fitment

If you have fitment problems, check sizing and bolt-hole patterns. Torque converters are sometimes mislabeled or misboxed to where they've no chance of fitting your application. They either won't fit the pilot or the transmission input shaft and stator. If a torque converter does not fit, do not force it.

8 Use Correct Flexplate Size

Most common flexplate sizes for C4 transmissions are 157-tooth (left) and 164-tooth (right). Each calls for its own bellhousing size. There's also a smaller 148-tooth flexplate (not shown here) for Mustang II and Capri.

9 Install Block Plate

This is a one-size-fits-all adjustable block plate for small-block Fords, available from Mike's Transmission. This block plate fits 157- and 164-tooth applications.

10 Use Transmission Fluid

Service the torque converter with 1 or 2 quarts of transmission fluid prior to converter installation. This primes the pump and gets prompt lubrication to transmission internals on start-up. You don't have to use Type F on old Ford transmissions anymore. Dexron III and Mercon IV are suitable fluids.

11 Lubricate Pump Drive

Converter pump drive should be lubricated with transmission lube prior to installation to prevent seal damage.

12 Install New Transmission Mount

Install a new transmission mount in order to minimize movement around the driveshaft's centerline. Also check driveshaft, universal joints, and slip yoke condition.

13 Adjust Shifter

Proper shifter adjustment is important to operation. Move the manual valve shift into Park; then, place shifter in Park. Secure linkage and tighten adjustment. Adjustment should fall in the middle (as shown).

14 Inspect Speedometer Drive Gears

Speedometer drive gears on transmission and cable should both be inspected for abnormal wear or damage. Install a new O-ring while you're at it and lubricate generously with transmission assembly lube.

15 Check Drain Plug Location

Critical Inspection

Torque converter drive stud and drain plug alignment are very important. Get this wrong and you wind up with a distorted flexplate. Make sure drain plug is located at the right hole and not up against the flexplate.

16 Run Transmission Lines

Transmission cooler lines should be hard lines between cooler/radiator and transmission. Keep the number of joints minimal.

Critical Inspection

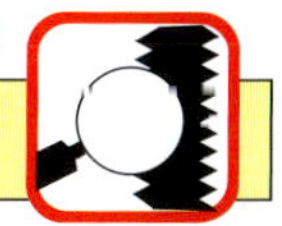

17 Inspect & Adjust Kickdown Linkage

Inspect and adjust the kickdown cable or linkage. Replace the neutral safety/back-up light switch if necessary. Check for proper operation and continuity.

18 Fill Fluid

Although vintage Ford automatics once called for Type-F fluid, they can get away with using Dexron III or Mercon IV these days. Type-F was more about friction enhancers than anything.

Professional Mechanic Tip

19 Don't Drain Torque Converter

Down the road, when the time comes for transmission service, never completely drain your torque converter for two reasons. First, you want fluid in the converter to prime the pump and ensure lubrication on start-up; second, some transmission professionals never completely change fluid due to the risk of "shocking" seals.

Troubleshooting Guide

It's good to have a source to guide you when you're struggling with troublesome transmission function. You can be methodical and cover all the bases during a rebuild and still things can go wrong. This is where a step-by-step approach to the problem can help get you steered in the right direction. As with just about any type of troubleshooting, begin with the simplest item first and work the problem from there. Don't panic, but assume the worst.

The first thing to do is find out whether the transmission works. With the engine running, does the transmission engage when you put it in drive or reverse? If not, immediately shut off the engine and check the transmission fluid level.

Is there fluid on the dipstick? If not, add fluid in small amounts until the dipstick shows at least "half." Then, start the engine.

You want a half-stick reading because fluid expands as it warms. Operating temperature is 150 to 170 degrees F. If (with a fluid reading on the dipstick) your transmission doesn't engage, control pressure likely isn't reaching servos and clutches. This means there's either no control pressure or pressure isn't making it to the clutches and servos. The absence of control pressure to components happens for several reasons: pump not making pressure, torque converter cavitated, stuck pressure relief valves, manual shift valve not working, valve body malfunction, failed servo piston, or failed clutch piston seals to name a few possibilities.

If there is fluid on the dipstick, check the manual shift valve, which is where control pressure begins.

Transmissions that are overfull tend to foam or aerate causing air bubbles to be drawn into the torque converter and pump, which causes a great disruption in control pressure. With the engine running, listen closely for abnormal noise such as torque converter improperly seated, which can cause squealing, no control pressure, and an inoperative transmission. An improperly seated torque converter can't turn the pump or the input shaft rendering a transmission inoperative. At the same time, it damages the pump, sending metal particles into the system doing further damage. When transmissions don't engage, most of the time it is an improperly seated torque converter, a pump issue, or a manual shift valve problem.

The following are common symptoms and their likely causes.

Transmission Doesn't Engage in Drive or Reverse

1. No control pressure.
2. Low fluid level.
3. Torque converter and/or front pump cavitated (no fluid or air in fluid).
4. Torque converter improperly installed, doesn't turn pump (listen for noise).
5. Manual shift linkage between car and transmission improperly adjusted.
6. Manual shift valve inside transmission pan not connected or stuck.
7. Pressure relief valve (or valves) unseated or missing.
8. Forward clutch pressure leak or improperly assembled (air check clutches by removing the valve body).
9. Blocked filter.

No 1-2 Upshift

1. Governor valves malfunction.
2. Governor improper installation on the output shaft (is the drive ball in place?)
3. Shift valve jammed.
4. Intermediate servo malfunction.
5. Broken or improperly assembled roller clutch.

No Upshift at Wide-Open Throttle

1. Vacuum modulator or rod malfunction
2. Governor or fluid passage malfunction.

Slippage in All Gears and Start Out

1. Low fluid level.
2. High fluid level (foaming).
3. Blocked filter.
4. Faulty vacuum modulator.
5. Insufficient pump pressure (sticking relief valve, gear/cavity damage).
6. Faulty torque converter.
7. Insufficient clutch engagement (pressure, clutch clearances).
8. Insufficient band engagement (improper adjustment, pressure).

No 1-2 Upshift

1. Governor issue, stuck valve.
2. 1-2 shift valve issue.
3. Intermediate band servo not applying.
4. One-way clutch defective or improperly assembled.

Slips at 2-3 Upshift

1. Improper direct clutch clearances.
2. Damaged direct clutch piston seal.
3. Leaking iron-sealing rings.

Early or Late 1-2 or 2-3 Upshift

1. Vacuum modulator or rod malfunction.
2. Valve body malfunction.
3. Governor malfunction.
4. Loose tubes on governor distributor.

Vehicle Slips or Ratchets in Park

1. Improper shifter adjustment.
2. Parking pawl damaged or improperly installed.

No Reverse

1. Low fluid level.
2. Manual shift valve out of adjustment.
3. Manual shift linkage improper installation and/or adjustment.
4. Vacuum modulator malfunction or failure.
5. Valve body assembly error.
6. Blocked filter.

No Engine Braking in Second Gear

1. Valve body issue.
2. Damaged intermediate band or improper adjustment.
3. Intermediate band servo piston seals.

Delayed Upshift or No Upshift When Cold

1. Hard or damaged clutch piston seals.
2. Insufficient pump pressure (sticking relief valve).

No Engine Braking in First Gear

1. Valve body issue.

Source Guide

B&M Racing & Performance
9142 Independence Ave.
Chatsworth, CA 91311
818/882-6422
www.bmracing.com

Leon's Transmission
7528 Reseda Blvd.
Reseda, CA 91335
818/345-8101
www.leanstransmission.com

Mike's Transmission
42541 6th St. East, #11
Lancaster, CA 93535
661/723-0081
www.mikestransmission.com

Performance Automatic
8174 Beechcraft Ave.
Gaithersburg, MD 28079
301/963-8078
www.performanceautomatic.com

Summit Racing
PO Box 909
Akron, OH 44398
800/230-3030
www.summitracing.com

TCI Automotive
151 Industrial Drive
Ashland, MS 38603
888/776-9824
662/224-8972
www.tciauto.com

Tom's Transmissions
16609 Sierra Hgwy
Canyon Country, CA 91351
661/251-3438
www.tomstrans.com

Transmission Rebuilding Company (TRC)
10140 Topanga Canyon Blvd.
Chatsworth, CA 91311
800/987-2676
818/882-0082
www.trctransmission.com

Transmission Parts & Cores
1981 West Winton Ave.
Hayward, CA 94545
510/783-5222
510/786-1162
www.transmissionpartsandcores.com

Transtar Industries, Inc.
7350 Young Dr.
Cleveland, OH 44146
800/359-3339
www.transtar1.com

TransGo
2621 Merced Ave.
El Monte, CA 91733-1997
626/443-7451
www.transgo.com